BIODIVERSITY
EXPLORING VALUES AND PRIORITIES IN CONSERVATION

DAN L. PERLMAN
GLENN ADELSON

Environmental Science and Public Policy
Harvard University
Cambridge, Massachusetts

**Blackwell
Science**

Blackwell Science Editorial offices:
Commerce Place, 350 Main Street,
  Malden, Massachusetts 02148, USA
Osney Mead, Oxford OX2 0El, England
25 John Street, London WC1N 2BL, England
23 Ainslie Place, Edinburgh EH3 6AJ, Scotland
54 University Street, Carlton, Victoria 3053, Australia
Other Editorial offices:
Blackwell Wissenschafts-Verlag GmbH Kurfürstendamm
  57, 10707 Berlin, Germany Zehetnergasse 6, A-1140
  Vienna, Austria

Distributors:
USA
Blackwell Science, Inc.
Commerce Place
350 Main Street
Malden, Massachusetts 02148
(Telephone orders: 800-215-1000 or 617-388-8250; fax
orders: 617-388-8270)

CANADA
Copp Clark, Ltd.
2775 Matheson Blvd. East
Mississauga, Ontario
Canada, L4W 4P7
(Telephone orders: 800-263-4374 or 905-238-6074)

AUSTRALIA
Blackwell Science Pty, Ltd.
54 University Street
Carlton, Victoria 3053
(Telephone orders: 03-9347-0300;
 fax orders 03-9349-3016)

OUTSIDE NORTH AMERICA AND AUSTRALIA
Blackwell Science, Ltd.
c/o Marston Book Services, Ltd.
P.O. Box 269
Abingdon
Oxon OX14 4YN
England
  (Telephone orders: 44-01235-465500;
   fax orders 44-01235-465555)

The Blackwell Science logo is a trade mark of Blackwell Science Ltd., registered at the United Kingdom Trade Marks Registry

Acquisitions: Jane Humphreys
Production: Lisa Flanagan
Illustrator: George Nichols
Cover design: Eliza K. Jewett
Manufacturing: Lisa Flanagan
Typeset by Northeastern Graphics Services Inc.
Printed and bound by Braun-Brumfield

Perlman, Dan L.
  Biodiversity : exploring values and priorities in conservation /
  Dan L. Perlman, Glenn Adelson.
    p.   cm.
  Includes bibliographiccal references and index.
  ISBN 0-86542-439-X (pb)
  1. Biological diversity. 2. Biological diversity
conservation.
  I. Adelson, Glenn. II. Title.
  QH541.15.B56P47 1997   97-11468
  333.95—dc21   CIP

Note about the cover: The man shown on the cover of this book had been a farmer until a few years before this photograph was taken. He had cleared his farm from what had been virgin rainforest. However, as a result of changes in his values, he left farming to become a forest guard, protecting against encroachment and cutting. One of the reasons we like this photograph is that it conveys some of the complexities in the relationship between humans and nature.

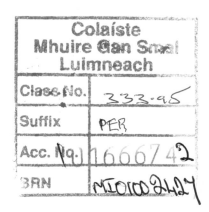

To Our Parents

Robert and Bernice Perlman
Lloyd Adelson and Iris Kaufman

Contents

Acknowledgments

This book owes it existence to two sets of people: those who made our course *Conservation Biology and Biodiversity* possible in the first place and those who helped this book grow from the course. We owe a great debt to all of the people involved in both of these endeavors.

Eric Fajer suggested that we take over the teaching of the Conservation Biology seminar at Harvard University that he taught so ably; without his enthusiasm and support, we would not have gotten so deeply involved in thinking about the topic of biodiversity. John de Cuevas's extraordinary personal support, and the financial support of the Baker Foundation, enabled us to create the field component that we believe is the key to our course. Woody Hastings, formerly head tutor in biology at Harvard, was invaluable in helping us fund and operate our course in the fashion we envisioned, and Ted Scovell has long offered generous support to the course. For the last few years Joan Donahue has given us unstinting help with the logistics of the class, and the Kann Rasmussen family and the Stone Foundation have generously supported the class financially during that same period. Our course assistants—Eve Kaplan, Charlotte Kaiser, Heather Leslie, Marco Simons, and Sara Goldhaber—made the course significantly better through their efforts.

On the book itself, we received invaluable comments on some or all of our manuscript from a number of people, including David Ackerly, Carol Bershad, Armando Carbonell, Brian Drayton, Les Kaufman, Kathleen Kelly, Drea Maier, Elaine Moise, Norman Myers, Mary Parkin, Dorothy Patent, Bernice Perlman, Judy Perlman, Jay Phelan, Bill Rice, Jeremy Rich, Ilene Rosin, Ted Scovell, and E. O. Wilson. Our students of the past few years have also read parts of the manuscript. All have offered us useful feedback during class discussions. We would especially like to thank the following students for their thoughts, which went beyond our in-class conversations, and for their help in performing library research: Roy Bahat, Ana Blohm, Luis Campos, Dan Cooper, Jamie DeNormandie, David Grewal, Eliza Jewett, Serena Kortepeter, Peter Kostishack, Rebecca Newton, Amy Salzhauer, Raphael Sperry, and Joshua Tosteson. We wish we could have incorporated even more of the wise comments that all of our readers offered.

Our deepest gratitude goes to the large number of people who supplied us, often on short notice, with crucial pieces of information: Jeanne Anderson, John Atwood, Henry Barbour, Fakhri Bazzaz, Joy Belsky, Grahame Bird, Richard Birndorf, John Cadle, Charles Cogbill, Gretchen Daily, Michael Donoghue, Lawrence Dowler, David Gewanter, William Haber, Michal Jasienski, Dick Johnson, Nels Johnson, Toby Kellogg, Robert Lawton, Olga Levaniouk, Richard Lewontin, Richard Lichtenstein, Trevor Lloyd Evans, Charles

Maier, Curt Meine, Luke Menand, Lenny Muellner, Norman Myers, Willa Nehlsen, David Olson, Mary Parkin, Mark Petrini, Zyg Plater, Martha Rojas, Peter Schoonmaker, Tim Simmons, Russ Spring Jr., Peter Stevens, Pat Swain, Margaret Symington, David Takacs, Sarolta Takacs, Fred Tauber, Peter Taylor, Peter Wayne, Alex Werth, Brendan Whittaker, Henry Woolsey, and Willow Zuchowski.

Our students Bella Sewall, Christina Carroll, and Eliza Jewett aided us in the final throes of pulling this book together; without their careful reading, the text would not be so polished and would probably contain large gaps. Eric Fajer helped us stay grounded in the real world with his thorough comments, and Bryan Norton gave two chapters a careful and thoughtful read at the last minute.

A few people deserve special recognition, for without their efforts this book and our experience in writing it would have been much poorer. Our editor, Jane Humphreys, has been deeply understanding and supportive, providing us with the latitude to explore many different ideas. Gregory Nagy has been a constant source of encouragement and a marvelously generous gateway into the world of the ancients; he has truly been our Mentor. Robert Perlman offered wisdom and constructive criticism at all stages of the writing and kept pushing us to improve the overall clarity of the book. We owe much of the quality of this book to these three people, and to the many others who have helped us develop the course and the book; whatever errors remain are ours alone.

Finally, we owe our deepest gratitude to our families. Nora Abrahamer and Jeremy Perlman (of the Perlman family) and Ilene Rosin, Ellery Rosin, and Bud Hanes (of the Adelson family) were all completely supportive of our efforts. Our wives, Nora and Ilene, gave us insight, criticism, and humor in equal measure; without their help we could not have written this book. We thank you all.

Foreword

Nearly a century and a half ago, Charles Darwin's *Origin of Species* gave the study of biological diversity a unifying explanatory framework: the theory of evolution by natural selection. Now the conservation movement has given it a unifying purpose. The salvaging of species of plants, animals, and microorganisms, and the genes that compose them, and finally of the ecosystems they themselves compose is now recognized around the world as a goal of highest urgency.

Scientists have come to understand that all of the Earth's environmental problems fall into two broad, fundamentally different categories. The first is deterioration of the physical environment, comprising toxic pollution, forced climatic warming, upper atmospheric ozone loss, and the depletion of natural resources. Most of these trends can be reversed and they will be, often at considerable cost. The second category is the reduction of biodiversity by extinction. It is mostly not reversible; when species and ecosystems in particular are lost, they cannot be replaced—ever. That is one way in which biodiversity reduction differs from purely physical change. The other is the great potential value that accrues when biodiversity is not only saved but scientifically studied and wisely managed for human benefit.

The study of biodiversity does not lend itself easily to abstract principles. It is best studied in the trenches, by people who care deeply about conservation, enjoy field research, and are drawn to the scientific concepts brought into play. This perspective is a principle merit of *Biodiversity: Exploring Values and Priorities in Conservation*. Dan Perlman and Glenn Adelson know first-hand which methods and principles work best. As seminar teachers and field leaders in biodiversity studies at Harvard University for the past several years, they understand the theoretical and practical problems most important to those learning the subject. As highly acclaimed teachers, they have demonstrated the ability to present their knowledge in a focused and effective manner. These desirable qualities stand out in *Biodiversity: Exploring Values and Priorities in Conservation*. It will serve as an excellent introduction and *vade mecum* for students, for decision makers, and in fact for any member of the public with a serious interest in conservation.

Edward O. Wilson

Legacies

The crane, returning every year, cries out
From the clouds above, and when you hear her voice,
Know that she means the time has come to plough,
The time of chilly rains. She gnaws the hearts
Of men who have no oxen.

—Hesiod, *Works and Days*[1]

So keep on looking for a bluebird
And list'ning for his song,
Whenever April showers come along.

—Buddy DeSylva, "April Showers"[2]

Imagine a cultural historian, perhaps a hundred years from now, who develops a database of the ways that birds have been used as symbols in art, music, and literature. Somewhere in this database she has stored the two lyrics excerpted above, one an evocative passage by one of history's great poets detailing the meaning of the crane to the farmers of ancient Greece, the other from a popular twentieth-century song by a Broadway lyricist. The Eurasian crane of Hesiod's great agricultural poem was a bird that symbolized foresight, justice, and retribution to the Greeks.[3] To twentieth-century North Americans, the eastern bluebird symbolizes hope and cheerfulness.

Our future cultural historian will probably not give a moment's thought to how it is that the lyrics of "April Showers" have survived for over a hundred years; instead, she will take for granted the powerful technologies that have given works of literature and music a form of immortality that could not have been imagined before the nineteenth century. The persistence of the work of Hesiod, on the other hand, was, throughout much of its history, not quite so secure. Some of the works of Hesiod, as well as those of Homer, the other great ancient Greek poet of the oral tradition, survive today, but historians tell us that the vast majority of Greek poetry, tragedies, comedies, and philosophy disappeared prior to the Renaissance. We have only scant references to large numbers of works and authors, perhaps a quoted sentence or two of an entire lifetime's production, that have otherwise vanished without a trace. For the classical literature that was saved, for what we have of Hesiod and Homer, of Plato and Aristotle, of Euripides, Sophocles, and Aeschylus, we owe an incalculable debt to scholars, librarians, and scribes who, over nearly 2000 years, preserved the ancient texts, copying them by hand onto papyrus and parchment. We know the names of only a few of the most prominent scholars and librarians of Alexandria and Byzantium who helped keep alive this rich cultural heritage; the others are nameless and faceless. These scholars, librarians, and scribes created a legacy as they fought off the ravages of time and war, copying the ancient texts from crumbling scrolls to fresh ones (see Figure 1-1).

The classical Greeks and other ancient peoples left us a literary legacy; they also left us agricultural and ecological legacies. Without the efforts of hundreds of generations of farmers from around the world, whose toil was detailed in all its difficulty by Hesiod, we would not have the highly domesticated grains and root crops that led to the development of permanent habitations and cities. Without cities, much of the world's culture as we know it today would not have developed.

The ecological legacy that we received from previous generations was not too different from the one they had received. As of the start of the twentieth century, most of the world's forests were intact, and most of the world's fauna and flora had survived the rise of human culture. The record, however, is not without exceptions: humans extirpated many of the bird species of the islands of the Pacific and Indian Oceans and nearly all of the large mammals of North America and Europe.* Deserts covered what was once the Fertile Crescent, and the formerly forested hills of Greece and Lebanon were largely treeless. Overall, though, most of the natural world had survived humankind's embrace.

From our vantage point in the present, we can look back at the legacies that our ecological forebears left us, just as we do with the librarians and scribes. What gaps in our ecological legacy do we long for the most? Just as we have scraps of text from lost tragedies and comedies that hint at what we have missed, we have fossils and—in the case of more recent extinctions—specimens, drawings, and written reports. The ecological legacy we received is missing some spectacular elements such as the saber-toothed cats, mastodons, woolly mammoths, giant ground sloths, and other large mammals of North America; the elephant bird of Madagascar; and perhaps 2000 species of indigenous birds on the Pacific islands[4] (see Figure 1-2). It was only by the late nineteenth and early twentieth century, as the billion-strong flocks of the passenger pigeon were being extinguished, that we begin to find analogs to the ancient librarians and scribes (see Color Plate 1). George Perkins Marsh, John Muir, Robert Porter Allen, Aldo Leopold, and Rachel Carson sounded early calls for the protection of our ecological legacy, and we owe a great debt to these visionary individuals and their colleagues, just as the whooping crane and bison, in one sense, owe their very existence to the labors of a few.

The level of human effort needed to preserve our cultural and ecological legacies has changed fundamentally in the last few hundred years, especially during this century. Until the invention of the printing press in the fifteenth century, works of literature had to be consciously and painstakingly copied by hand. Without the application of human planning and effort, not a single

---

*There remains some question whether the Pleistocene megafauna were extirpated by humans or by climatic change, or by a combination of those and other factors. See P. S. Martin and R. G. Klein, *Quaternary Extinctions* (Tucson: University of Arizona Press, 1984), for details.

Figure 1-1.

Hesiod's *Works and Days.* The text of this poem, which began as part of the oral tradition, is a legacy passed down to us through the efforts of many. Once the poem was written down, scribes had to copy the text painstakingly from crumbling papyrus, as seen in this figure, to fresh papyrus or parchment. At times of crisis, this work was chosen by librarians and scribes to be saved, while other works of antiquity were lost. (Reproduced from Aloisius Rzach, *Hesiodi Carmina* [1902] Lipsiae in Aedibus B. G. Teubneri)

Figure 1-2.

The saber-toothed cat (*Smilodon*). One of the spectacular elements missing from our ecological legacy, the saber-toothed cat went extinct, along with most other large mammals of the Western Hemisphere, about 9000 to 12,000 years ago.

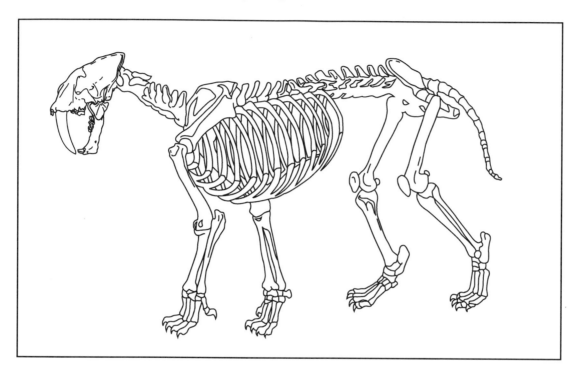

copy of any of the great literary works of classical Greece would have reached us. Texts were mortal and fragile. At any moment, a work only existed on a few pieces of papyrus; a small number of unfortunate events such as fires, leaky roofs, or invasions could destroy all existing copies of a text, thereby destroying the work itself. In contrast, elements of biodiversity such as genes, species, and ecosystems were relatively untouched by humans. With a few exceptions, such elements could survive in much the same manner as they had before humans evolved and spread to cover the planet.

Today, individual works of literature and music have near immortality. The music and lyrics of "April Showers" are printed in thousands of songbooks; recorded on many thousands of records, CDs, and tapes; and stored on numerous computer disks. The song is so widely distributed that it would require vast numbers of individual events of destruction to seriously threaten the existence of the work. Thus, no more can we reserve the appellation "immortal" for only the greatest works of art, such as the poetry of Hesiod, for today's technology assures that all forms of popular art have become virtually immortal. Unfortunately, many elements of our planet's ecological legacy are headed down the opposite path. The explosive growth of the human population squeezes plant and animal populations ever more tightly. Fewer "copies" of most species and ecosystems exist; many of those that still exist are in ever more dangerous situations as our species extends its impact to all ecosystems.

Humanity today is in the position of the librarians and scribes of Alexandria and Byzantium. We know that the actions we undertake now, as well as those that we do not, will determine the ecological legacy that we leave for generations to come. We know that the task requires difficult choices, increased knowledge, careful planning, and concentrated effort. If we do not make the most of our moment, if we do not do the difficult work of protecting our ecological legacy, the cultural historian we met at the beginning of the chapter may not know the crane and the bluebird as living species. Instead, she may know them only as historical artifacts and symbols used by humans in literature and art, like the dodo or the dinosaurs. The Eurasian crane (*Grus grus*) of Hesiod's *Works and Days* experienced a severe population crash, mostly due to the draining of wetlands throughout Europe beginning about 300 years ago. By the eighteenth century, nesting cranes could no longer be found in Greece, although they still could be seen there in migratory flight. Only through the work of dedicated conservationists over the last 50 years have some Eurasian crane populations begun to rebound, although in many parts of their range they are still declining, and they are not known to have recovered their breeding grounds in Greece.[5] Additionally, the eastern bluebird of North America (*Sialia sialis*) experienced severe population declines during the middle of the twentieth century due to destruction of habitat, expansion of urbanization, a few very cold winters in the 1950s, and the increase in population of introduced competitors, such as the starling and the house sparrow. Only through concentrated conservation efforts, most particularly artificial nest-box projects, have we begun to reverse this decline.[6] Conservationists have just begun the fight against the loss of the world's ecological legacy, joining the ranks of others, such as the ancient and medieval librarians, who long ago began the fight against the loss of our literary legacy.

OF CRITICS, CRITERIA, AND CRISES

Along with the literary and ecological legacies that we received from the ancients, we also inherited a rich linguistic legacy, which includes some words germane to our discussions of biodiversity.* The Greek word for "decide" was *krínein*, from which we get *critic* (via the noun *krités*, "to judge," and *kritikós*, "able to make judgments"), *criterion* (a standard for making a judgment), and even *crisis* (via the noun, *krísis*, "judgment").[7] From this single linguistic root, we can sketch out the key features of the ancient librarians' jobs, as they attempted to preserve their cultural legacy.[8]

The librarians of antiquity functioned as critics and employed criteria in selecting texts to copy and thus preserve. The librarians had to choose which texts to copy first, given the limited lifespan of handwritten copies on papyrus. Furthermore, scholars often had to make selections among varying

*This discussion was suggested to us by Gregory Nagy of the Harvard University Department of the Classics and we are deeply indebted to him for sharing his ideas and his knowledge.

versions of a text, choosing the version that, in their opinion, was the closest to the text as originally written by the author, or was the version that they thought should be preserved for other, often political, reasons.

This critical selection process was especially important during most of the Middle Ages, when the libraries of Byzantium were the sole repository for much of Greek literature. Given the high costs of copying a single text, and the limited resources of Byzantium, the librarians had to actively set priorities of what to preserve; they simply could not decide to preserve all texts. The issue of limited resources would become even more pressing during times of crisis, as when fire, flood, or invaders threatened the physical existence of the libraries. At these times, the librarians and their assistants had to select a few critical texts to carry to safety.

Today we face another crisis—the rapid loss of the Earth's biological diversity. Tropical deforestation is causing an untold number of species extinctions; exotic pests are easily transported to almost any spot on the globe, where they can have devastating effects on local species and ecosystems; and genes that might hold cures for cancers or AIDS are disappearing before we even know of their existence. To what extent are we thinking critically, and employing clearly spelled-out criteria, as we craft our response to this crisis? To date, two basic strategies have been employed in biodiversity conservation.

- The first is responding to the "degree of threat," choosing those species and ecosystems in gravest danger of immediate extinction or irreversible damage and directing resources to stave off those threats. This is the approach of the U.S. Endangered Species Act.

- The second is attempting to protect as many species as possible in order to maximize the probability of protecting biodiversity that one values. A count of the number of species in a region, or "species richness," is the most often used method.

While both of these guidelines have served well as a first response and should be part of a range of strategies available to conservation decision makers, both have fundamental problems. We are convinced that our generation can preserve our ecological legacy better, by using a broader range of criteria that are more specifically tailored to the various goals we have in protecting biodiversity.

The first of the criteria we highlighted above, degree of threat, has one serious drawback. Using it as the primary guide for setting priorities strips conservation decision makers of most of their ability to work purposefully toward a future they want. They function much the way that our health care system might if all resources were allocated to emergency room doctors and none to preventative medicine. Most of the conservation resources available get expended in reacting to crises as they arise rather than planning for our world's future good health. This emphasis on degree of threat may result from the assumption that we should, and can, protect all elements of biodiversity (i.e., all genes, species, and ecosystems). If we attempt to protect all of biodiversity, then it makes sense to use all available resources for saving the most threatened elements. If, instead, we recognize that we cannot protect all, then we should choose intelligently among a wide variety of criteria and priorities, of which threat is just one. As economists David Pearce and Dominic Moran have written:

> The reality is that little can be done to prevent huge increases in the world's population—it is in that respect "too late" for a good deal of the world's biological diversity. If so, it is essential to choose between different areas of policy intervention—not everything can be saved. This view is reinforced by the fact that the world is extremely unlikely to devote major resources to biodiversity conservation.[9]

The second of the highlighted criteria, using species richness as the primary guide to conservation decision making (i.e., attempting to protect as many species as possible), also suffers from serious drawbacks. It is a reasonable strategy in the absence of specific knowledge about the elements of biodiversity under consideration for protection. However, in cases in which we *do* have specific information about the biology of the species and ecosystems, such a strategy may fail to target the elements of biodiversity that we value most. People may value species that are endangered rather than common; those that play key roles in their ecosystems; those that are endemic (restricted in distribution) to a region; those that contain potentially valuable chemical compounds; those that, like the crane and bluebird, have symbolic worth; and those that have no close relations among other extant species. Decision makers frequently have knowledge, and are able to use that knowledge to effectively target elements of biodiversity, to place resources where they will do the most good, rather than relying on the species richness approach, which takes none of the important particulars into account.

There will be times when protecting the most endangered element of biodiversity will not be the wisest criterion; there will be times when it makes more sense to protect an area with fewer species than an area with more species—if we can better satisfy specific goals that way. In writing this book, we hope to help conservationists begin to enumerate the full range of criteria they can use to target elements of biodiversity for protection and to determine which criteria can best be employed in which cases. But in order to proceed intelligently, we must first try to make as rigorous a statement as possible of what biodiversity is. Most commentators, logically enough, begin by proffering a definition of the term.

DEFINING BIODIVERSITY

A review of the current literature could leave one with the impression that a reasonable consensus exists on the meaning of the term *biodiversity*. Most definitions of the term refer to genes, species, and ecosystems, as this quote from the 1992 compendium, *Global Biodiversity*, makes clear: "It has become a widespread practice to define biodiversity in terms of *genes, species,* and *ecosystems*, corresponding to three fundamental and hierarchically related levels of biological organization."[10] Box 1-1 provides further definitions of *biodiversity* and *biological diversity* that all use more or less the same language. But on closer examination, such definitions turn out to be of limited use in practice. As the authors of *Global Biodiversity* also note: "The term 'biodiver-

BOX 1-1. *Several definitions of the cognate terms* biological diversity *and* biodiversity *show strong similarities.*

"Biological diversity refers to the variety and variability among living organisms and the ecological complexes in which they occur. Diversity can be defined as the number of different items and their relative frequency. For biological diversity, these items are organized at many levels, ranging from complete ecosystems to the chemical structures that are the molecular basis of heredity. Thus, the term encompasses different ecosystems, species, genes, and their relative abundance."[12] —Office of Technological Assessment, 1987 [We take up this definition in more detail at the beginning of Chapter 2.]

"'Biological diversity' means the variability among living organisms from all sources including, inter alia, terrestrial, marine and other aquatic ecosystems and the ecological complexes of which they are part; this includes diversity within species, between species and of ecosystems."[13] —Convention on Biological Diversity, 1992

"*Biological diversity* (= Biodiversity). Full range of variety and variability within and among living organisms, their associations, and habitat-oriented ecological complexes. Term encompasses ecosystem, species, and landscape as well as intraspecific (genetic) levels of diversity."[14] —Fiedler and Jain, 1992 (glossary entry)

"Biodiversity is the total variety of life on earth. It includes all genes, species and ecosystems and the ecological processes of which they are part."[15] —International Council for Bird Preservation, 1992

"*Biodiversity* The variety of living organisms considered at all levels, from genetics through species, to higher taxonomic levels, and including the variety of habitats and ecosystems."[16] —Meffe and Carroll, 1994 (glossary entry)

"Definitions of biodiversity usually go one step beyond the obvious—the diversity of life—and define biodiversity as the diversity of life in all its forms, and at all levels of organization. . . . 'All levels of organization' indicates that biodiversity refers to the diversity of genes and ecosystems, as well as species diversity."[17] —Hunter, 1996

sity' is indeed commonly used to describe the number, variety, and variability of living organisms. This very broad usage, embracing many different parameters, is essentially a synonym of 'Life on Earth.'"[11]

Elements of biodiversity,[18] those entities that conservationists study and protect, can be envisioned on many different levels, from alleles to the entire biosphere, as cataloged in Figure 1-3. Despite the wide array of levels at which we could analyze biodiversity, the definitions of biodiversity that most conservationists employ refer only to ecosystems, species, and genes. One test of whether such definitions are sufficient would be to ask, Can we use these three terms in practice to assess the biodiversity of a region?

If one takes the standard definition of biodiversity as including all genes, species, and ecosystems, one must first determine what ecosystems and species are present; we devote most of Chapter 6 to exploring how involved this task is, as the terms ecosystem and species are fraught with difficulty. Even if one could overcome the conceptual difficulties with these terms, the number of microorganisms, for example, at any site that are as yet unknown to science is staggering.[19] Furthermore, the problem of where to draw boundaries between ecosystems is as dependent on the needs of the investigator as on the biology of the area.[20] Next, one would have to ascertain what

Figure 1-3.

Elements of biodiversity. If biodiversity is defined as all of "Life on Earth," then one can describe biodiversity, and list the elements of biodiversity, on many levels. These range from the level of alleles, of which there are many trillion, to the level of the biosphere, of which we have only one. By convention, most definitions list ecosystems, species, and genes as the elements of biodiversity, although one could just as easily list landscapes, populations, and alleles.

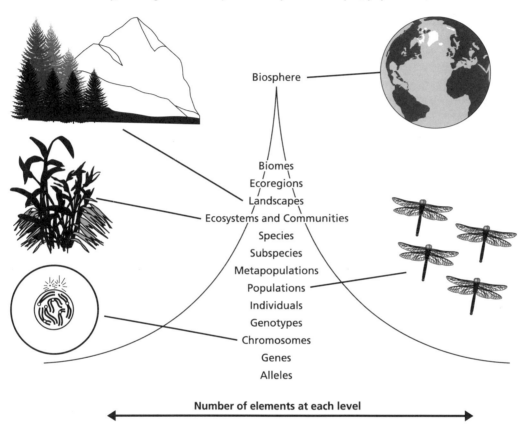

Biosphere

Biomes
Ecoregions
Landscapes
Ecosystems and Communities
Species
Subspecies
Metapopulations
Populations
Individuals
Genotypes
Chromosomes
Genes
Alleles

Number of elements at each level

genes are present. The U.S. National Institutes of Health are currently funding an enormously costly undertaking, the Human Genome Project, to catalog the genes of the human species. It is expected to take 15 years and cost $3 billion to determine the genetic makeup of just a single species.[21] We know virtually nothing of the genetics of the vast majority of the other species on Earth. Each of these tasks in itself is monumental; when taken together, the enterprise becomes virtually impossible. If biodiversity is taken to mean the entirety of genes, species, and ecosystems within the area of interest, it is currently no more possible to assess the biodiversity of a cubic meter of forest than to assess the biodiversity of the entire planet.

The current definitions of biodiversity as "genes, species, and ecosystems" fail both in theory and in practice. First, they do not recognize the conceptual difficulties inherent in the constituent terms of biodiversity (namely, genes, species, and ecosystems). Second, they ignore the practical and technical problems involved in making real-world inventories of biodiversity.

Third, they fail to take into account the incommensurabilities between different levels—how does one equate species with ecosystems, for example, in determining the biodiversity of an area? Finally, these definitions make no distinction in the worth of elements of biodiversity within any given level, as in the problem of species richness that we described above.

To protect biodiversity, we need to act. In order to act, in order to answer the question "If we want to protect biodiversity, what should we do?" we must have a concept of biodiversity that can guide our actions. The massive complexity of biodiversity "out there," in the "real world" prevents us from being able to catalog, understand, or protect it in its entirety. When called upon to decide whether it would be better, for the protection of biodiversity, to dedicate our limited resources to the captive breeding of the California condor, or to the preservation of the Ecuadorian Amazon, or to the eradication of exotic weeds such as purple loosestrife, a definition of biodiversity that is equivalent to "Life on Earth," or "all genes, species, and ecosystems," is of no help at all. We must work with more restricted practical definitions. Accordingly, the term biodiversity, from this point on in the book, will refer only to the human conceptions of biodiversity that allow us to narrow down the intractable Life on Earth to scales that allow us to solve specific conservation problems. It will not refer to the objective, unknowable biodiversity out there in the real world, which will be referred to as Life on Earth.

In our experience, when conservation decision makers use the term biodiversity they are not referring to all of Life on Earth, rather they are referring to a specific subset for a specific purpose. That is, they are employing their own conception of what we are calling biodiversity in this book, and that conception is shaped by their values, biases, and interests. Conservation decision makers need to be aware of this process and of the array of more tractable concepts that allows them to tailor their actions toward specifically articulated goals. The criteria that we presented earlier—the degree-of-threat criterion and the species richness criterion—are two important and useful concepts, but as we discuss, each can be counterproductive when it is used as the sole or primary criterion for conservation decisions. Decision makers have many other criteria available for setting priorities among biodiversity conservation projects, and these criteria are based on the diverse values that they hold.

Central to this book is the assertion that real-life assessments and evaluations of biodiversity are neither totally objective nor totally scientific. Once biodiversity is thought of in this way in practice, we can see it as a value-laden set of concepts that is delimited by the ideals, goals, biases, and interests of those employing the term. Moreover, we believe that once conservationists begin to recognize the diversity of human conceptions of biodiversity, they will realize that this is a real strength rather than a failing.

Diversity, then, is a part of every facet of this discussion. There is a wonderful diversity of plants, animals, and microorganisms "out there" that has captured the imagination of so many people. There is the added diversity of ecosystems and ecological interactions—pollination, predation, symbiosis—which are equally important aspects of biodiversity that are often more difficult to categorize. But there is also the remarkable diversity of human beings and human cultures, each with their own complex view of what all this biological diversity means to them. Conservationist Reed Noss highlighted this idea in the first line of a seminal 1990 paper: "Biological diversity (biodiversity) means different things to different people."[22] As a result of this insight Noss proposed a rich characterization, instead of a single definition, for the term *biodiversity*. Similarly, as we discuss throughout this book, no single person or viewpoint captures the complete richness of either the concept of biodiversity or the actuality of Life on Earth. It is only with the values and views of many different biologists, farmers, social scientists, politicians, business people, chemists, artists, native healers, and others that we can truly appreciate all that biodiversity is and all that biodiversity can offer to humans.

BIODIVERSITY IS A CONCEPT THAT SHOULD BE EXPLAINED AND NOT DEFINED

Biodiversity is a term that represents a complex concept. It resists definition for two reasons. The first, which we just illustrated, is that its meaning depends on the values of the people who use it. The second is that it represents underlying entities that are, themselves, everchanging. In his introduction to *Keywords: A Vocabulary of Culture and Society*, British literary critic Raymond Williams says of terms that represent complex concepts: "Variations and confusions of meaning are not just faults in the system, or errors of feedback, or deficiencies of education, they are in many cases . . . substance."[23] Of the keyword "nature" he notes: "Nature is perhaps the most complex word in the language . . . any full history of the uses of nature would be a history of a large part of human thought." When we begin to think about nature we see that the fundamental signal of nature is change, that is, evolution. And therefore it is unreasonable to expect that we can unambiguously fix a single definition on a term whose purpose is to describe something that is fundamentally in a state of constant and unpredictable change. So it is with the term biodiversity, which encompasses elements that are undergoing evolution. Change is necessarily built into any concept of biodiversity: species evolve, ecosystems undergo succession, genes mutate.

Most of the literature on biodiversity does not define it in a way that it can be employed in real-life assessments; practical definitions are either ignored, assumed, or explicitly stated without rigorous analysis. Upon reflection, this phenomenon is not surprising. Consider the opening line of the Declaration of Independence of the United States: "We hold these truths to be self-evident, that all men are created equal, and are endowed by their Creator with certain unalienable rights, that among these are Life, Liberty, and the Pursuit of Happiness." Clearly there are many terms in this sentence capable of various interpretations, none of which is rigorously defined in the document that follows. Particularly of interest are the terms *equal* and *rights*. No other two terms have been as important in the history of the United States, yet these are not simple terms that can be unambiguously defined in a single paragraph. Ronald Dworkin, in *Taking Rights Seriously*, and Amartya Sen, in *Inequality Reexamined*, are just two of the many scholars to devote entire treatises to the assessment of these terms.[24] Even the definition of the word *men* as used in this sentence, which would appear simple and unambiguous, has changed drastically along with society. In one sense, we could view the collected opinions of the U.S. Supreme Court as answers to the questions, "What is the definition of 'rights'?" and "How do you assess 'equality'?"

The founders of the United States held the concept that "all men are created equal" to be a basic tenet, a fundamental axiom. Since those words were written, the task of interpreting and putting into practice the meanings of those terms has fallen to the jurists of the United States. In a landmark case in 1896, *Plessy v. Ferguson*, the Supreme Court ruled that "separate" railcars designated for black and white citizens were to be considered "equal" under the Constitution. Only in 1954 did the Court change its interpretation on the subject; in the case of *Brown v. Board of Education of Topeka, Kansas* the court unanimously ruled that "separate but equal" was not the same as "equal," thereby changing the legal definition of the word *equal*.

Our approach in writing this book is to treat the term *biodiversity* similarly to *rights* and *equality*—first, as a broad term that describes a general good, and second, as one that is not susceptible to a single or a simple practical definition; instead, it is a term whose meaning changes as the values of its users change. The task that awaits conservation decision makers, and that we hope this book will help begin, is the process of developing an appreciation for the complexities of biodiversity. The terms *equality* and *rights* have had to be repeatedly redefined over the course of history, and we believe the same will be true for *biodiversity*.

ANOTHER LEGACY AND THE ORIGINS OF THIS BOOK

For the past seven years, the authors have been teaching a class called *Conservation Biology and Biodiversity* to undergraduates at Harvard College. The class lasts for a full academic year and includes three weeks of field work in both New England and Costa Rica. Our conviction in developing the class was that only extensive field experience would enable our students to fully understand the conceptual issues in this discipline. The field experience allows our students to compare firsthand the biodiversity of different ecosystems by looking carefully at the organisms and processes within them. Moreover, the students meet and learn from a wide variety of people involved in conservation decision making, including conservation biologists, government policy makers, developers, farmers, legal scholars, and grassroots conservation activists (see Figure 1-4).

The class is our attempt to pass important lessons to the generation that will soon be taking on important roles in conservation planning and action; it is our attempt to help protect the planet's legacy of biodiversity. Many of our field experiences have allowed our students to learn lessons that they would not have otherwise learned, and we have found some of these experiences to be so valuable that we have chosen them to illustrate key concepts and issues throughout this book. In an ideal world, we would be able to share them directly with you, the reader, to bring you to the sites we visit, to show you the organisms we study, to introduce you to our guest speakers. Deep down, we believe that the most effective way to understand the importance and wonder of biodiversity is to go into nature and experience the magic of

Figure 1-4.
The Conservation Biology and Biodiversity class. Students meet informally with (A) former Secretary of the Environment of Vermont, Bren Whittaker, (Photograph by Dan L. Perlman) and (B) former Minister of Energy, Natural Resources, and Mines of Costa Rica, Alvaro Umaña (left), at his butterfly farm. (Photograph by Glenn Adelson)

A

B

a damselfly nymph molting or the splendor of an old-growth northern hardwoods forest in autumn, as we do each year with our students. Of course, we cannot do this literally with each of our readers; we hope, however, that the vignettes from our class that we include here help to bring to life the challenging and sometimes difficult concepts and issues that arise throughout this book.

We see you, our readers, as conservation decision makers, a group that includes conservation biologists, lawmakers and policy planners in governmental and nongovernmental agencies, and business people in fields such as agriculture, forestry, and paper products. Students and other individuals who can live their lives in ways that help protect our biodiversity legacy are also important conservation decision makers. Some of these people will someday be in positions where they can make conservation decisions that will affect biodiversity in important ways, but even those who are not can put the knowledge they gain from this book to good use. Moreover, this book is not aimed only at people in developed nations. Although we recognize the unlikelihood of including them in our actual readership, rural people in developing countries are among the world's most directly involved conservation decision makers, and we must keep in mind their values, needs, and activities whenever the term *conservation decision maker* or a discussion of values arises in this book.

ROADMAP TO THE REST OF THIS BOOK

In this chapter we introduced the issues that we will deal with in the rest of the book. In Chapter 2, "Conceptual Problems with Biodiversity," we explore the term *biodiversity* in depth and demonstrate some of the subtleties and difficulties associated with the term.

Chapter 3, "The Role of Values," begins our discussion of the role that human values have to play in defining and assessing biodiversity. To date, when the words *biodiversity* and *value* have been linked, they have typically been considered in light of the value of biodiversity for humans; in other words, the worth of biodiversity for humans. In Chapter 3 we discuss the relationship between values (or beliefs and preferences) and worth (or value for humans), particularly as these concepts pertain to the conservation of biodiversity, and review several classifications of values regarding biodiversity and worth of biodiversity.

Exploring the meaning of the term *diversity* is the subject of Chapter 4, "Diversity." We discuss how practical definitions of the term vary from person to person. Applying this understanding to biodiversity, we find that practical definitions of the term are based on human values, interests, and preferences, since these cause individuals to focus on certain elements of biodiversity, while excluding others from consideration.

We consider the distribution of biodiversity in Chapter 5, "Mapping the Patterns of Biodiversity." We introduce three "landscapes": the geographic or physical landscape, in which we consider the distribution of biodiversity around the planet; the temporal landscape, in which we consider changes to biodiversity through time; and the taxonomic landscape, in which we discuss the distribution of biodiversity in various taxonomic groups. In Chapter 5 we point out that different research techniques offer vastly different views of biodiversity and that one's choice of technique will shape one's views of biodiversity.

In Chapter 6, "Ambiguities," we revisit usage of the term *biodiversity* and of the terms that are often seen as its component parts, namely, *species, ecosystems,* and *genes.* Each of these terms is ambiguous in a variety of ways, both in nature and in the way that humans speak and write. In Chapter 6 we describe several ambiguities, and how these ambiguities can cause problems for decision makers using the term *biodiversity*.

Understanding how values shape field inventories, a basic component of biodiversity decision making, is the subject of Chapter 7, "Inventories." Although seemingly a strictly objective technical exercise, the biodiversity inventory is in many ways quite subjective. Since it is impossible to catalog the biodiversity of a region exhaustively, inventory takers must decide what aspects of the region to inventory and which to leave aside.

We consider the process of setting priorities for biodiversity conservation in Chapter 8, "Articulating Goals and Setting Priorities." In large part, this chapter serves to integrate and sum up the content of the previous chapters, and explores the implications that various human values and goals have on the process of setting biodiversity conservation priorities. This chapter highlights the effects that values have on all aspects of the priority-setting process.

Finally, in Chapter 9, "Future Directions," we look toward the future and suggest that conservation decision makers can turn the value-laden essence of biodiversity assessment to their advantage. In our view, every bit of human insight that can be brought to bear on the issue of setting priorities for biodiversity conservation is important, and an understanding of human values regarding biodiversity is perhaps our most useful guide for setting priorities.

WHAT THIS BOOK IS AND WHAT IT IS NOT

This book is about biodiversity—but not in the way you might expect. Books about biodiversity tend to fall into one of four categories. First, they can be exhortations to protect the planet's ecological legacy, ably done in works such as *The Diversity of Life* by Edward O. Wilson and *The Primary Source* by Norman Myers. Second, they can be encyclopedic treatises that attempt to catalog the species and ecosystems of the world, such as *Global Biodiversity*. Third, they can be books that discuss the methodological tools for protection, such as *Global Biodiversity Assessment*. Finally, many books, especially conservation textbooks like Hunter's *Fundamentals of Conservation Biology*, Primack's *Essentials of Conservation Biology*, and Mette and Carroll's *Principles of Conservation Biology*, combine aspects of all three of these categories.[25]

This book, on the other hand, is not so much about the actual biological entities out in the real world but rather about the ways in which humans conceive of biodiversity, assess biodiversity, and set priorities for conserving biodiversity. Each of the processes of conceptualizing, assessing, and setting priorities calls on the technical skills of biologists such as ecologists and systematists (those who study the evolutionary relationships between species), yet the processes are, in our view, more than narrowly defined biological problems; they are broad-based human enterprises with a large social and political component. Today, the general public greatly values the ecological legacy that our generation has received, as evidenced by the level of interest in nature and wildlife. In addition, conservation biologists and policy makers have been developing effective techniques for protecting bio-

Three steps of biodiversity conservation.

To better understand the niche of this book, try to visualize the conservation of biodiversity as a three-step process, the first and third of which have been well addressed by the books mentioned above. This book focuses on step 2.

- Step 1: *Exhortation.* Learn to love and appreciate nature, which leads to a desire to protect biodiversity.

- Step 2: *Articulation.* Articulate what you value in nature and which elements of biodiversity you want to target for protection.
- Step 3: *Protection.* Protect the elements of biodiversity that you have targeted.

diversity. What is missing is a middle step: the articulation of what we value and how those values affect what should be targeted for protection (see Box 1-2).

Life on Earth is so rich and complex that the human mind cannot encompass it in its entirety and human technology cannot calculate its myriad components. As a result of this richness, humans necessarily focus on subsets of Life on Earth—in our terminology—and use them as surrogates for the whole of biodiversity in their assessments. Moreover, we are convinced that human resources are not sufficient to protect all of Life on Earth and that difficult choices of what to conserve lie ahead. It is our goal in this book to make clear the role that human values play in creating practical definitions of the term *biodiversity*, in assessing biodiversity, and in setting priorities for conserving biodiversity. We hope that this book can serve as a guide to thinking rigorously about the wonders of biodiversity.

References

1. Hesiod, *Works and Days*, in *Hesiod and Theognis*, trans. Dorothea Wender (London: Penguin Books, 1973), lines 449–454.
2. B. G. DeSylva (words) and Louis Silvers (music), "April Showers" (Miami, Florida: Warner Bros., 1921).
3. John Pollard, *Bird in Greek Life and Myth* (Plymouth, England: Thames & Hudson, 1977), 96–98; Paul Johnsgard, *Cranes of the World* (Bloomington: Indiana University Press, 1983), 70–71; Gregory Nagy, "Theognis of Megara: A Poet's Vision of his City," in *Theognis of Megara: Poetry and the Polis*, ed. T. J. Figueira and G. Nagy (Baltimore, Maryland: Johns Hopkins University Press, 1985).
4. David Steadman, "Human-caused Extinction of Birds," in *Biodiversity II*, ed M. L. Reake-Kudle, D.E. Wilson, and E.O. Wilson (Washington, D.C.: Joseph Henry Press 1997).
5. Hartwig Prange, "Crane," in *Birds in Europe: Their Conservation Status*, ed. G. M. Tucker and M. F. Heath

(Cambridge: BirdLife International, 1994), 234–235.
6. See J.M. Speirs, *Birds of Ontario* (Toronto: Natural Heritage/Natural History, 1985), 612; L. Zeleny, *The Bluebird—How You Can Help Its Fight for Survival* (Bloomington, Indiana: Indiana University Press, 1976).
7. John Ayto, *Dictionary of Word Origins* (New York: Arcade Publishing, 1990).
8. Many of the details were supplied in personal communications by Dr. Gregory Nagy and Dr. Sarolta Takacs of Harvard University, as well as from Gregory Nagy, *Pindar's Homer*, rev. paperback ed. (Baltimore, Maryland: Johns Hopkins Press, 1994), 61–62.
9. David Pearce and Dominic Moran, *The Economic Value of Biodiversity* (London: Earthscan Publications, 1994), 32.
10. World Conservation Monitoring Centre, *Global Biodiversity: Status of the Earth's Living Resources* (London: Chapman & Hall, 1992), xiii.

11. World Conservation Monitoring Centre, *Global Biodiversity: Status of the Earth's Living Resources* (London: Chapman & Hall, 1992), xiii.
12. U.S. Office of Technological Assessment, *Technologies to Maintain Biological Diversity* (1987).
13. Convention on Biological Diversity, Article 2 (1992).
14. Peggy L. Fiedler and Subodh K. Jain, eds., *Conservation Biology: The Theory and Practice of Nature Conservation Preservation and Management* (New York: Chapman & Hall, 1992), 484.
15. International Council for Bird Preservation, *Putting Biodiversity on the Map: Priority Areas for Global Conservation* (Cambridge: International Council for Bird Preservation, 1992), 3.
16. Gary K. Meffe and C. Ronald Carroll, *Principles of Conservation Biology* (Sunderland, Massachusetts: Sinauer, 1994), 559.
17. Malcolm L. Hunter Jr., *Fundamentals of Conservation Biology* (Cambridge, Massachusetts: Blackwell Science, 1996), 19.
18. Walter V. Reid and Kenton R. Miller, *Keeping Options Alive* (Washington, D.C.: World Resources Institute, 1989).
19. Carol K. Yoon, "Counting Creatures Great and Small," *Science* 260 (1993): 620–622.
20. Gordon Orians, "Endangered at What Level?" *Ecological Applications* 3, no. 2 (1993): 206–208.
21. Office of Technology Assessment, *Federal Technology Transfer and the Human Genome Project* (1995).
22. R. F. Noss, "Indicators for Monitoring Biodiversity: A Hierarchical Approach," *Conservation Biology* 4, no. 4 (1990): 355–364.
23. R. Williams, *Keywords: A Vocabulary of Culture and Society* (New York: Oxford University Press, 1976).
24. R. M. Dworkin, *Taking Rights Seriously* (Cambridge, Massachusetts: Harvard University Press, 1977); A. K. Sen, *Inequality Re-examined* (Cambridge, Massachusetts: Harvard University Press, 1992).
25. Edward O. Wilson, *The Diversity of Life* (Cambridge, Massachusetts: Harvard University Press, 1992); Norman Myers, *The Primary Source* (New York: W. W. Norton, 1984); World Conservation Monitoring Centre, *Global Biodiversity: Status of the Earth's Living Resources* (London: Chapman & Hall, 1992); United Nations Environment Programme, *Global Biodiversity Assessment* (Cambridge: Cambridge University Press, 1995); Malcolm L. Hunter Jr., *Fundamentals of Conservation Biology* (Cambridge, Massachusetts: Blackwell Science, 1996); Gary K. Meffe and C. Ronald Carroll, *Principles of Conservation Biology* (Sunderland, Massachusetts: Sinauer, 1994); Richard B. Primack, *Essentials of Conservation Biology* (Sunderland, Massachusetts: Sinauer, 1993).

2

Conceptual Problems with Biodiversity

Conscious of the intrinsic value
of biological diversity and of the
ecological, genetic, social, economic, scientific,
educational, cultural, recreational and aesthetic values
of biological diversity and its components . . .
Determined to conserve and sustainably use
biological diversity for the benefit of
present and future generations . . . [1]

— Opening lines of the United
Nations Convention on Biological
Diversity

In June 1992, the city of Rio de Janeiro hosted the largest gathering of heads of state ever to occur in the history of the Earth. The United Nations Conference on Environment and Development, known popularly as the Earth Summit, attracted leaders of more than 170 countries.[2] During the conference, representatives from 158 of the world's nations signed the Convention on Biological Diversity, and since then 166 nations have ratified the Convention.[3] (George Bush, President of the United States, refused to sign the Convention at the conference. Although his successor, Bill Clinton, eventually signed the Convention, the U.S. Senate has yet to ratify it.)[4] The Convention, whose opening words began this chapter, was nearly four years in the writing.[5]

Despite its lofty role as the subject of a major international convention, the term *biological diversity* is, in its current sense, remarkably young. According to biologists J. L. Harper and D. L. Hawksworth, the meaning of the term began to approach its current usage only in 1980, and only started being used in the sense applied in the Convention in 1986, just six years before the Convention was signed (see Figure 2-1).[6] Here we continue the task that we began in Chapter 1 by taking an even closer look at this term. Throughout this book we will treat the term *biological diversity* and its commonly used contraction, *biodiversity*, as interchangeable, given the history that we relate below and their indistinguishable definitions.

Figure 2-1.

Increase in usage of the term *biodiversity* in the literature. Harper and Hawksworth found a dramatic increase in the number of times the term *biodiversity* was found in the scientific literature between 1988 and 1994. (After J. L. Harper and D. L. Hawksworth, "Preface to Biodiversity: Measurement and Estimation," *Philosophical Transactions of the Royal Society of London* 345 [1994]: 6)

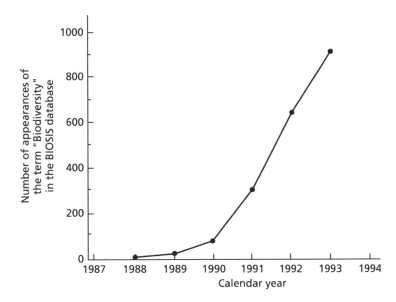

AN EARLY DEFINITION OF THE TERM BIODIVERSITY

Let us expand on our discussion of the definitions of biodiversity offered in Chapter 1 by taking a close look at one of the earliest and most detailed of those definitions. The term *biodiversity* was probably first used by Walter G. Rosen while organizing the conference that came to be known as the National Forum on BioDiversity, held in Washington, D.C., during September 1986.[7] Selected papers from that forum were eventually published in the volume *BioDiversity*, edited by Edward O. Wilson.[8] In 1987, the U.S. Office of Technological Assessment (OTA), at the request of several U.S. House and Senate committees, issued a report entitled *Technologies to Maintain Biological Diversity*. The OTA volume begins with a box entitled "What Is Biological Diversity?" which we have reproduced here as Box 2-1.[9] Rosen has stated that he felt that the definition from the 1987 OTA report captured what he had in mind when he came up with the term *biodiversity*.[10]

The OTA definition explicitly articulates some of the various levels (ecosystems, species, and genes) at which biological diversity can manifest itself. Moreover, the discussion implies that the concept *biological diversity* is easily understood, largely because the examples provided present no conceptual difficulties. Many different definitions of biodiversity have been written since the publication of this OTA definition, and the vast majority of them

BOX 2-1.

An early definition of the term biological diversity, *from* Technologies to Maintain Biological Diversity, *U.S. Office of Technological Assessment, 1987. We include this definition not because we endorse it, but to use it as a starting point for discussions about the terms* biological diversity *and* biodiversity.

What Is Biological Diversity?

Biological diversity refers to the variety and variability among living organisms and the ecological complexes in which they occur. Diversity can be defined as the number of different items and their relative frequency. For biological diversity, these items are organized at many levels, ranging from complete ecosystems to the chemical structures that are the molecular basis of heredity. Thus, the term encompasses different ecosystems, species, genes, and their relative abundance.

How does diversity vary within ecosystem, species, and genetic levels? For example,

- *Ecosystem diversity:* A landscape interspersed with croplands, grasslands, and woodlands has more diversity than a landscape with most of the woodlands converted to grasslands and croplands.

- *Species diversity:* A rangeland with 100 species of annual and perennial grasses and shrubs has more diversity than the same rangeland

after heavy grazing has eliminated or greatly reduced the frequency of the perennial grass species.

- *Genetic diversity:* Economically useful crops are developed from wild plants by selecting valuable inheritable characteristics. Thus, many wild ancestor plants contain genes not found in today's crop plants. An environment that includes both the domestic varieties of a crop (such as corn) and the crop's wild ancestors has more diversity than an environment with wild ancestors eliminated to make way for domestic crops.

Concerns over the loss of biological diversity to date have been defined almost exclusively in terms of species extinction. Although extinction is perhaps the most dramatic aspect of the problem, it is by no means the whole problem. The consequence is a distorted definition of the problem, which fails to account for many of the interests concerned and may misdirect how concerns should be addressed.

are merely restatements or slight refinements of the statement above (a few of these definitions can be found in Box 1-1). Most definitions refer to multiple levels of biodiversity, and most give no indication that the assessment of biodiversity is difficult.

Let us consider the difficulties inherent in the term *biodiversity*, as well as some of the concepts that we believe can help clarify the usage of the term. We offer four examples that illustrate the subtleties and conceptual difficulties of the term *biodiversity*.

THE PROBLEM OF CONSIDERING MULTIPLE LEVELS OF BIODIVERSITY SIMULTANEOUSLY

Consider the following statement from the OTA definition of biological diversity: "A landscape interspersed with croplands, grasslands, and woodlands has more diversity than a landscape with most of the woodlands converted to grasslands and croplands." The truth of this statement might seem self-evident; however, the statement is not universally true. The particulars of the landscapes under consideration must be examined to decide in which cases the statement is indeed true.

Imagine the Midwest of North America as it existed until the mid-nineteenth century. When European settlers first reached this region, they found large expanses of tall-grass prairie mingled with oak savanna—an area full of grasslands and woodlands, as in the OTA description. Within a few short decades, however, much of the region was "converted to grasslands and croplands," including wheat and corn (maize) fields, vegetable gardens, and pastures of European grasses for forage (see Figure 2-2). But was there a loss of biodiversity?

On this landscape, at the level of *ecosystem* diversity, there were actually more ecosystems after the settlers' initial plantings than before the arrival of the settlers: patches of the original two ecosystems (prairie and savanna) were mixed in with patches of wheat and corn fields, gardens, and pastures. (We recognize that the definition of the term *ecosystem* is quite difficult, and we return to this issue in Chapter 6.) So, in direct contrast to the OTA statement, the landscape with most of the original grasslands and woodlands converted to grasslands and croplands actually had more biological diversity—at the level of ecosystems—than the original landscape.

There are several objections one might pose to this example, such as, "Who cares about the biodiversity in wheat and corn fields, vegetable gardens, and pastures? They're human-imposed, unnatural ecosystems, and what's more, the world is filled with fields and gardens that look just like these." These objections have a good deal of merit but are implicitly based on certain values, albeit ones held by almost all conservationists. These values are: 1) that human artifacts should not be included in assessments of biodiversity[11] and 2) that common elements of biodiversity are less valuable than uncommon ones.[12] One might also object that, after some of the natural ecosystems have been converted to managed ones, the landscape would have less biodiversity because the wheat and corn fields, gardens, and pastures together would not support as many species as the landscape would gain from having the prairie and savanna back—but such an objection fails on two counts. First, this objection is pitched at the level of *species* diversity and does not include ecosystem diversity in the assessment, and both levels are valid and important components of biodiversity, according to the OTA definition. Second, the objection may be wrong factually; studies have shown that a disturbed woodland often has more species than an undisturbed one.[13] This example highlights the importance of tracking the level of biodiversity that one has specified, because levels are incommensurable with each other—that is, they cannot be expressed in terms of a common unit. In the next example we begin to address the question of whether biodiversity can be adequately assessed quantitatively, that is, whether more elements of biodiversity are necessarily "better" than fewer elements of biodiversity.

Figure 2-2.

Grasslands, woodlands, croplands, and rangelands. When the tall-grass prarie of North America was converted from its presettlement state of (A) grasslands and woodlands to (B) grasslands, woodlands, rangelands, and croplands, did ecosystem diversity go up or down? Did biodiversity go up or down?

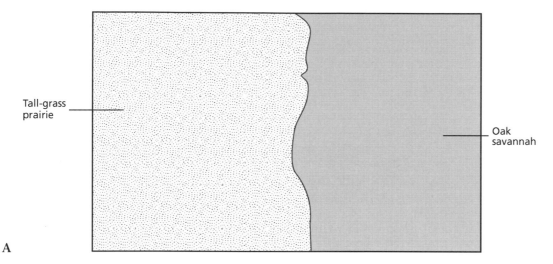

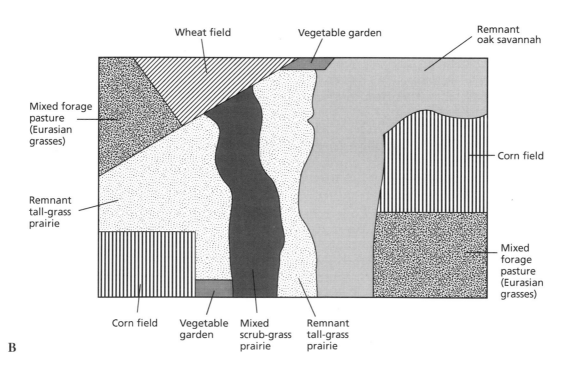

If Biodiversity Is Good, Is More Biodiversity Always Better than Less Biodiversity?

Every year during our class trip to Costa Rica we take our students to the tiny community of Monteverde, about 35 kilometers up a rocky dirt road from the Pan American Highway. Two kilometers from the center of the community stands the entrance to the privately owned Monteverde Cloud Forest Preserve. As our students enter the preserve, they are invariably struck by the variety of the plant life. Along a typical five-meter section of the path, we see several species of shrubs in each of four plant families: the coffee family, the African violet family, the black pepper family, and the melastome family (a common family of the New World tropics). Patches like this may also contain species of begonia, heliconia (a relative of the bird-of-paradise flower), and lobelia, with its beautiful, elongated, hummingbird-pollinated flowers. Up above, in the tree branches, sit numerous bromeliads, in the pineapple family, and perhaps some orchids. Farther up in the canopy are the leaves of the trees themselves, while on the forest floor grow myriad species of herbs and mosses and ferns. Connecting all these different layers of the forest are stringy vines and thick, woody lianas (see Color Plate 2).

At certain spots in the preserve, especially along the Sendero El Camino, the central trail, we see a single species of pink-flowered plant that stands out as both beautiful and abundant; it is known locally as "La China" and scientifically as *Impatiens wallerana* (see Figure 2-3). Abundant as it is, the impatiens is not native to Costa Rica. If we had visited this forest 200 years ago, we would not have seen this plant; it was brought to the region from Southeast Asia for ornamental gardens, after which it escaped and became naturalized in the forest.

The impatiens has probably not caused any native plants to go locally extinct, as it grows among widespread, weedy, second-growth plants. Its presence has therefore increased the overall number of plant species in the Monteverde Cloud Forest Preserve by one. By the most commonly used measure of biodiversity, the simple species count (known technically as species richness), a forest with the impatiens contains more biodiversity than

Figure 2-3.
Impatiens wallerana. This beautiful flower, native to Asia, found its way into Costa Rican rainforests by escaping from ornamental gardens. In what ways has the addition of this species increased the biodiversity of the Monteverde Cloud Forest Preserve? (Photograph by Dan L. Perlman) (See Color Plate 3)

would the same forest without the impatiens. Similarly, according to more complex measures of biodiversity, such as those that take evolutionary relationships among species into account, the forest of today contains more biological diversity than previously. Not only is the impatiens new to Monteverde, but the entire impatients family is new to the area around the preserve. So we have added not only one plant *species* but also one plant *family* to the region.

But if we canvassed our students, or a group of conservation decision makers (including biologists) visiting the preserve, what would they say about the lovely impatiens? When we have asked this question, we have found that opinions on the subject are deeply divided. Many conservation decision makers prefer to see the impatiens removed from the forest, for a variety of reasons. Some state that the presence of an exotic species reduces the integrity of the ecosystem in some fashion,[14] while others fear that the increase in species numbers is a temporary illusion and that the introduction of an exotic species will surely create a loss of native species over time. Still others react on the basis of aesthetics, reasoning that the impatiens simply does not "belong" in this forest (any more than the Australian eucalyptus trees that are planted throughout Costa Rica belong there), while others worry that the presence of the exotic species might lead them to draw improper conclusions about the evolutionary and ecological forces that shaped the ecosystem over evolutionary time. In contrast, some biologists point out that the plants do little damage to the functioning of the forest, as they occur largely in second-growth areas, and that they are not worth the effort it would take to eradicate them. But we doubt that a single conservation decision maker would actively applaud the addition of this lovely little flower, and none would suggest bringing in other exotic species in order to "increase" biodiversity. Let us examine this situation further.

Conservation biologists, to consider one set of viewpoints in particular, are deeply in favor of biodiversity. Because the impatiens adds a new species, and even a new family, to the ecosystem isn't there more biodiversity present in the preserve than before? And isn't that good? In response to these two questions, most would answer, "Ah, hmm, it depends . . . " or "No." Most conservation biologists acknowledge that biodiversity is not a straightforward concept and that although species richness is often used in biodiversity assessments, it is not the whole story. In practice, not all elements of biodiversity are valued equally; in contrast, species richness as a measure of biodiversity *does* value all species equally. And the impatiens, in Monteverde, is an element of biodiversity that is not highly valued by conservation biologists, despite its pretty flowers. As for whether the impatiens makes a positive contribution to the biodiversity of the Monteverde preserve, most conservation biologists agree that the preserve's biodiversity, or at least its integrity, is *diminished* with the arrival of the impatiens. No conservation decision maker would disvalue the impatiens as an element of Life on Earth; neither would any conservation decision maker disvalue the impatients as an element of biodiversity *in its native habitat* in southeast Asia. It is only when humans enable this species to invade other habitats that many conservation decision makers would claim that it should be eradicated.[15]

Throughout the book we continue addressing the issue of when an element is considered a positive contribution to the biodiversity of an assemblage and when it is not, as this is not a problem that can be answered simply or quickly. In the next example we explore further the proposition that conservation decision makers do not value all elements of biodiversity equally. Just as the impatiens added an entirely new family to Monteverde, the subject of our next example adds a great deal to our planet's biodiversity as the sole representative of its entire taxonomic order.

DOES MORE DIFFERENT MEAN MORE IMPORTANT?

The tuatara (*Sphenodon punctatus*) is a single species of meter-long, superficially iguana-like reptile that resides on approximately 30 islands surrounding the North Island of New Zealand (see Figure 2-4).* It is the last of the Sphenodontida, an ancient order that reached its peak diversity in the Triassic, some 200 million years ago. Humans know little of the biology of this long-lived, slow-moving, nocturnal animal, because it is so difficult to observe—the tuatara lives in the isolation of uninhabited, cliff-bound, windswept islands surrounded by heavy seas. But the tuatara is isolated in another sense—in a taxonomic, evolutionary sense. All the other members of its order are extinct; its closest living relatives are the 6000 species of lizards and snakes that constitute the order Squamata (see Figure 2-5).

Conservationists, and biologists in general, have always shown a special concern for species that are taxonomically isolated; species that are the "last of a line," or otherwise have no close relatives, are the last living repositories of information otherwise lost in deep evolutionary time.[16] It is not necessarily easy, however, to answer the question of how to value the tuatara relative to another species. Is the tuatara any more deserving of conservation efforts than the green iguana (*Iguana iguana*), which was formerly widespread throughout much of the New World tropics but is now considered endangered due to habitat destruction and overhunting?

One method of quantifying the worth of the tuatara would be to include it as just one more species in a species count. But in doing so, the tuatara would be accorded no special worth at all. An island with populations of the tuatara and 12 nonrelict vertebrate species would have no more biodiversity, under this measure, than an island with populations of 13 nonrelict vertebrate species that did not have the tuatara on it.

---

*The taxonomy of the tuatara is problematic—there is some evidence that a single population of the species (*Sphenodon punctatus*) should be segregated into a separate species, *Sphenodon guntheri*. See C. H. Daugherty, A. Cree, J. M. Hay, and M. B. Thompson (1990): "Neglected Taxonomy and Continuing Extinctions of Tuatara (*Sphenodon*)," *Nature* 247 177–179. For the purposes of this discussion, we will treat all of the extant tuataras as a single species, *Sphenodon punctatus*.

Figure 2-4.

Tuatara and green iguana. These two reptiles are superficially similar, but all of the species related to the tuatara (A) went extinct approximately 200 million years ago (R.M. May. "Taxonomy as Destiny" *Nature* 347 [1990]: 129–130). On the other hand, the green iguana (B) is in the same family as more than 900 other species of iguanas (G.R. Zug [1993] *Herpetology*). (Photographs by Mark Moffett/Minden Pictures)

A

B

Figure 2-5.
Cladogram of the reptiles. The tuatara is the sole representative of the order Sphenodontida; lizards and snakes taken together make up the order Squamata. The green iguana from Figure 2-4B is one of many species in the family represented by the branch labeled New world iguanas. The entire suborder Serpentes (snakes) is collapsed into one branch. The other branches represent selected families of lizards. (The term *reptile*, as used here, specifically excludes turtles and their relatives and crocodiles and their relatives.) (Adapted from G. R. Zug, *Herpetology* [San Diego, California: Academic Press, 1993])

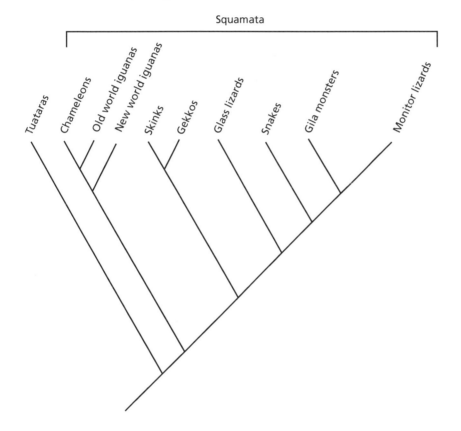

A second method for assessing biodiversity depends upon the position of the species in its diagram of evolutionary relatedness, such as the one shown in Figure 2-5 (these are commonly known as phylogenetic trees, or cladograms*), to arrive at a value for that species relative to the other species in that diagram. The question has often been put in terms of conservation effort or conservation expenditures: If we are faced with the current endangerment of more species than we have time or money to protect, how do we go about deciding which of the currently endangered species to devote our

*A phylogenetic tree, or cladogram, is a graphical representation of speciation events through evolutionary time and the relationships between terminal taxa (which may be species, but can also be higher taxa, such as the orders and families represented here, or lower taxa, such as subspecies). For a more complete explanation, please see Chapter 5. The cladograms presented in this book are based on our reviews of the literature and on discussions with colleagues, rather than on original research by the authors.

efforts to and which regretfully to ignore or underfund? Conservationists have developed an array of complicated methods for analyzing relative evolutionary isolatedness in order to arrive at assessments of relative value, which may help us deploy our conservation resources.

How much more effort should we put into protecting the tuatara than the iguana? Should we value them equally? After all, the tuatara and the green iguana are each one species. Or should we value the tuatara as equal to all 6000 species of the Squamata combined? Since both the Squamata and the Sphenodontida are orders, and the tuatara represents the entirety of the order Sphenodontida, perhaps we should give the two orders equal weight. Or is there an intermediate way of valuing the tuatara relative to the iguana? And if so, how do we create a sensible method of determining this value? There is no one right answer; this topic has been a subject of considerable debate, with several groups of researchers proposing methods for weighing the relative worth of different taxa.[17]

THE WORTH OF AN ELEMENT OF BIODIVERSITY DEPENDS ON CONTEXT

Our final example demonstrates that evaluations of biodiversity are highly context specific, that even though an element of biodiversity does not change, our assessment of its worth can change markedly—if the world around it changes. A striking example of this phenomenon can be seen in one of the most widely planted crops of the modern world, corn (*Zea mays*, or maize as it is known outside of the United States). By the late 1960s, plant breeders had achieved remarkable sophistication in what they had done to this crop, which is used for human food and animal feed, and as a precursor of chemical processes such as production of ethanol. Over 99% of the corn crop in 1969 was grown from specially produced hybrid seed, according to corn genetics expert Paul C. Mangelsdorf.[18] Among the greatest achievements by breeders had been the discovery of a strain of corn from Texas that produced sterile "male" flowers, thus obviating the need for cutting the male "tassels" from each growing plant, a formerly labor-intensive and expensive process. (According to Mangelsdorf, there was a time in the United States when "on the peak day of the season some 125,000 persons in the United States were engaged in removing tassels from corn plants."[19]) The feature of male sterility was considered so valuable that by 1969, some 70% to 90% of the U.S. hybrid corn crop carried this trait.

In 1970, however, a particularly nasty form of the southern corn-leaf blight, caused by a mutation in the fungus *Helminthosporium maydis*, appeared in the United States—and plants that carried the Texas trait of sterile male flowers were especially susceptible to the blight. Total corn production was down by 15% that year, with some states experiencing much larger drops in production; the cost of the lost corn was $1 billion.[20] Luckily, plant disease specialists were able to isolate the problem quickly, and breeders came up with a solution. This example demonstrates that the evaluation of a particular element of biodiversity, in this case the gene for male sterility, depends on the context in which the evaluation is made. The gene was considered invaluable until the mutant form of blight appeared, at which point the worth of this particular gene dropped precipitously.

RETHINKING THE ASSESSMENT OF BIODIVERSITY BY INCLUDING CONTEXT AND PARTICULARS

As we have heard over and over again, tropical rainforests are treasure troves of the world's biodiversity, containing tens to hundreds of times more species of plants, birds, and insects than forests of equivalent size in the north temperate zone. Biologists have found over 600 species of butterflies in the Monteverde region alone, an area smaller than Rhode Island; this compares with the roughly 440 species of butterflies known from eastern North America (see Figure 2-6).[21] But perhaps the common image of rainforest biodiversity, exemplified by butterfly species diversity, has dulled our analytical approach to answering the question: How can we make precise assessments of biodiversity? The statement that the plant or butterfly biodiversity of a specific tropical rainforest is "greater" than that of almost any temperate ecosystem is true on many levels, across many definitions, regardless of whether we assess biodiversity in terms of species, genes, ecosystems, or something else, or even any combination of terms. But that does not mean that the biodiversity of butterflies in Monteverde is necessarily of higher conservation priority than the butterfly biodiversity of Rhode Island. One might find reasons why the Rhode Island butterfly fauna is especially important, such as if the Rhode Island fauna has a high proportion of endangered species or a high proportion of taxonomically isolated species. Furthermore, the more subtle comparisons make the assessment of biodiversity especially difficult: assessing the addition of human-created ecosystems to a landscape, determining the effect of the impatiens on the biodiversity of Monteverde, or weighing the importance and worth of the tuatara or an individual gene in hybrid corn.

Two pairs of interlocking components must, we believe, be part of any effective assessment of biodiversity priorities: 1) particulars and context and 2) values and worth. The next chapter is devoted to an exploration of the roles that values and worth play in biodiversity assessment, so we will not discuss them in detail in this chapter. In this section we highlight the importance of facts about elements of biodiversity and the setting in which these facts are considered; in other words, what we call *particulars* and *context*.

In most situations, observers pay special attention to the particulars, the details, of specific elements of biodiversity. Because the impatiens was exotic and was brought to Central America by humans, most conservation biologists would evaluate its impact differently from that of another exotic species, the cattle egret, a bird that arrived in the New World under its own power and reached Costa Rica in 1954.[22] Similarly, the gene that created male sterility in hybrid corn had tremendous economic importance only because of its particular effect on the corn phenotype; had the gene instead created a tiny difference in yield or improved protein content slightly, it would not have had the same importance.

Figure 2-6.

A comparison of Rhode Island and Monteverde. Although considerably smaller than Rhode Island, the Monteverde region contains over 600 species of butterflies; all of eastern North America has only 440 species of butterflies.

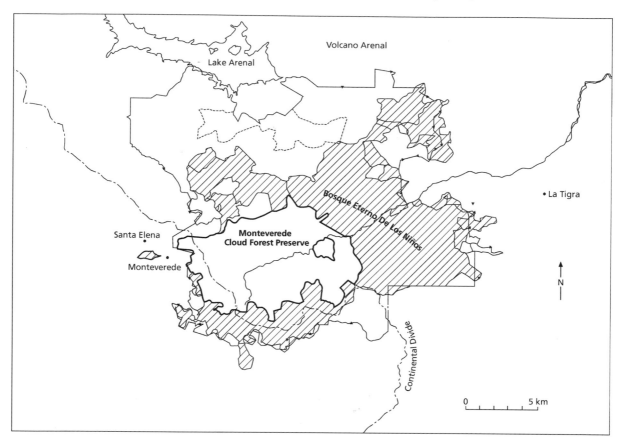

Assessments of a particular feature depend greatly on the context in which that feature is being evaluated. For example, the Texas male sterility gene acquired great worth and became widespread only because its context included the following crucial factors:

- Hybrid corn seed produces a much higher yield than nonhybrid corn seed.

- The production of hybrid seed requires the detasseling of growing corn.

- No machine exists to detassel corn.

- Detasseling by hand is very labor intensive.

- Labor in the United States is quite expensive.

In this context, the particulars of the male sterility gene had tremendous economic worth, as it could replace the expensive labor of human detasselers. Had any aspects of the context been different, the Texas male sterility gene would not have been deemed so valuable and would not have been used so

Figure 2-6.
(Continued)

widely, and the 1970 corn blight would not have been so severe. In fact, several years before the Texas male sterility gene was discovered, a different male sterility gene was found, but this gene was never used widely. As it happened, the non-Texas sterility gene was tightly linked with a gene that produced white corn; since farmers of that time greatly preferred yellow corn, the gene was never employed in production.[23]

Similarly, the wheat and corn fields in the OTA example received little respect in terms of what they add to the biodiversity of a landscape largely because of their context, namely, that vast numbers of wheat and corn fields exist in the world. If, instead, only one wheat field and one corn field were left, then the importance of these fields—specifically for the biodiversity they contain—would be enormous.

INTERACTIONS AMONG CONTEXT, PARTICULARS, VALUES, AND WORTH

Human values largely determine which particulars and contexts one considers to be of great worth. Concepts of aesthetics, for example, cause one to search for novel shapes and striking color combinations, while a desire for good health has been the motivator for humans trying thousands of plants as remedies for various ailments. In contrast, few people place any value on the patterns of wing venation found on insects (although taxonomists often place great emphasis on such features for distinguishing closely related species), and few actively consider wing venation as either a particular data point or useful context in any conservation decision (see Figure 2-7).

Particulars and context together form the factual basis for setting priorities in biodiversity conservation. We need facts—particulars—about specific elements of biodiversity in order to make informed assessments of biodiversity. But isolated facts alone, without context, are of little use. If a conservation decision maker only knows the fact that a certain uncommon plant species produces a certain useful chemical compound, he or she cannot adequately assess the conservation importance of the species. Perhaps five other species, each widespread and abundant, produce the same chemical, or chemists may be able to synthesize the compound cheaply and in large quantities. In either case, the addition of context may drastically alter the priority that the conservation decision maker places on the uncommon species in question.

Particulars and context cannot be completely separated from each other. What is a particular when considering a certain element of biodiversity becomes an aspect of context when a different element is considered. Together, these facts-on-the-ground create the arena in which values get put into action. Indeed, they help to shape human values, by defining the realm of the possible and defining what is rare and what is common, and it is to values that we turn in the next chapter (see Figure 2-8).

EXHORTATIONS

Our goal in writing this book is to help the reader understand better the real-world implications and problems raised by exhortations to protect biodiversity. We believe that the concept of biodiversity is both more subtle and more difficult to employ in real-world situations than is commonly appreciated.

One of the key issues facing those who would protect biodiversity in today's world can be illuminated with a brief retelling of the biblical story of Noah and the Flood. The God of the Old Testament initially intended to wipe all living things from the face of the Earth, but after being persuaded not to do so, told Noah (in the language of the day) to protect biological diversity. In many ways, humankind is at a comparable stage today—we have clear injunctions from international conventions such as the Rio Conference and national legislation such as the Endangered Species Act, from impassioned conservationists and from various religious traditions, to protect our biodiversity leg-

Figure 2-7.
Wing venation. This seemingly trivial particular can be important to taxonomists for distinguishing species—although most people never consider patterns of wing venation in different insects. Here you can compare the wing venation of the dragonfly *Ladona julia* (A) with that of the related *Libellula incesta* (B). Many dragonfly workers place the members of the genus *Ladona* in *Libellula*. (from J. G. Needham and M. J. Westfall, Jr., *A Manual of the Dragonflies of North America* [Berkeley: University of California Press, 1954] 473, 478)

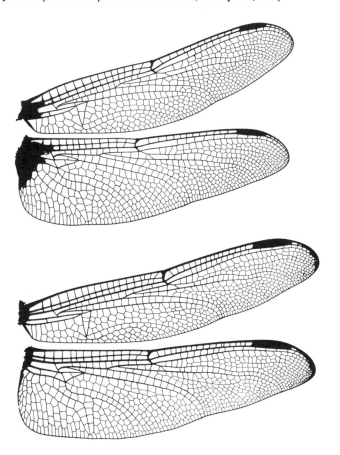

acy. A crucial difference between Noah's situation and ours is that although the underlying goals that we are enjoined to follow are very similar, Noah received clearly spelled-out objectives and strategies. All he had to do was build an ark 300 cubits long by 50 cubits wide by 30 cubits high, get a male and female of each animal onto the ark by a certain deadline, and his task was complete.* In our time, we are still fumbling about, trying to understand what type of ark to

---

*Noah was actually commanded to load not one, but *seven* of each sex for "clean beasts" and "fowl of the air"—indicating perhaps that God valued certain species more than others (Genesis 7:2–3).

create and how we should stock it. This book attempts to help the reader think about how to go about answering these questions.

Yet another difference between Noah's situation and the situation of today is that, according to the Bible, Noah was able to save representatives of every kind of living creature on the planet—and that task is not feasible today. The limited finances and limited effort available for biodiversity con-servation, combined with the growing human population and growing ex-

---

Figure 2-8.

Context and particulars. The entities that we call *context* and *particulars* are all facts-that-ex-ist-in-the-world, yet individual actors treat different facts as context and others as particulars. Particulars can be seen as those aspects of context that an individual can affect in some way or in which he or she is most interested. For instance, a corn breeder, an inventor, and the president of a multinational farming corporation might consider the efficient production of corn quite differently and probably view different facts as the particulars over which they have control.

For the corn breeder, the fact that a genetic trait exists to create corn with sterile male flow-ers is the fact with particular importance, for this is something that the breeder can act upon to change the overall situation. The inventor sees the opportunity to create a detasseling ma-chine as of particular interest. The president of the farming corporation could view the fact that labor in the United States is expensive as the most important particular—for the corpora-tion could choose to move its corn production to another country where labor is cheaper.

Together, all of the facts form a context—but each actor selects certain particulars on which to focus and act. Each actor's particular is part of the other actors' context.

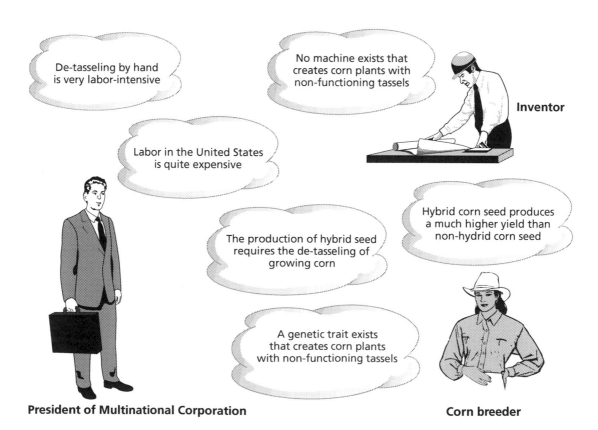

pectations of living standards, render the task of protecting all of biodiversity virtually impossible. Between the time we write these words and the time you read them, many habitats around the world will have been greatly reduced in area, and several species, perhaps a great many, will have gone extinct. Humankind cannot protect all of the species and habitats extant at any given time. But as a species, a species with unprecedented power over the planet, we humans can make wise decisions about where to begin the task of protecting elements of biodiversity.

We recognize that many of our students and many lovers of nature would prefer nothing more than to live in a world in which nature is able to take care of itself, in which we could leave ecosystems alone and they would return to their preindustrial state. In all but the most remote areas of the planet, we can no longer afford the luxury of that ideal. The workings of nature have been severely disrupted, and many of the parts necessary to return it to its previous state are either missing, as are wolves in most temperate forests, or gone forever, as is the passenger pigeon. We write this book convinced not only of the necessity of human choice in conservation policy but also of the central role that human values will play in these choices. We write this book for those who will have to make the difficult decisions that determine what of nature will remain.

REFERENCES

1. Convention on Biological Diversity.
2. Timothy E. Wirth, "The Road from Rio—Defining a New World Order," *Colorado Journal of International Environmental Law and Policy* 4 (1993): 37–44.
3. http://www.unep.ch/bio/ratif.html on 13 June 1997; and World Resources Institute, *World Resources 1994–95* (Oxford: Oxford University Press, 1994), 154.
4. http://www. unep.ch/bio/ratif.html on 13 June 1997.
5. http://www.iisd.ca/linkages/vol09/0918001e.html on 5 December 1996.
6. J. L. Harper and D. L. Hawksworth, "Biodiversity: Measurement and Estimation," *Phil. Trans. R. Soc. Lond. B* 345 (1994): 5–12.
7. J. L. Harper and D. L. Hawksworth, "Biodiversity: Measurement and Estimation," *Phil. Trans. R. Soc. Lond. B* 345 (1994): 5–12.
8. E. O. Wilson, ed., *BioDiversity* (Washington, D.C.: National Academy Press, 1988).
9. U.S. Office of Technological Assessment, *Technologies to Maintain Biological Diversity* (1987).
10. Walter G. Rosen, personal communication, July 1994.
11. Paul L. Angermeier, "Does Biodiversity Include Artificial Diversity?" *Conservation Biology* 8 (1994): 600–602. But for an alternative viewpoint, see Thomas Palmer, "The Case for Human Beings," *The Atlantic Monthly* 269 (1992): 83–88.
12. For example, see Malcolm L. Hunter Jr., "The Mismeasure of Biodiversity," in *Fundamentals of Conservation Biology* (Cambridge, Massachusetts: Blackwell Science, 1996), 23–25.
13. J. F. Franklin and R. T. T. Forman, "Creating Landscape Patterns by Forest Cutting: Ecological Consequences and Principles," *Landscape Ecology* 1 (1987): 5–18.
14. Paul L. Angermeier and James R. Karr, "Biological Integrity Versus Biological Diversity as Policy Directives," *Bioscience* 44 (1994): 690–697.
15. For example, see P. L. Angermeier, "Does Biodiversity Include Artificial Diversity?" *Conservation Biology* 8 (1994): 600–602.
16. See R. M. May, "Taxonomy as Destiny," *Nature* 347 (1991): 129–130 Martin Weitzman, "What to Preserve: An Application of Diversity Theory to Crane Conservation," *Quarterly Journal of Economics* 108, no. 1 (1993): 157–184; P. H. Williams, R. I. Vane-Wright, and C. J. Humphries, "Measuring Biodiversity for Choosing Conservation Areas," in *Hymenoptera and Biodiversity*, ed. John LaSalle and Ian D. Gauld (Wallingford, United Kingdom: CAB International, 1993), 310–311. But see T. L. Erwin, "An Evolutionary Basis for Conservation Strategies," *Science* 253 (1991): 750–752.
17. For example, P. H. Williams, R. I. Vane-Wright, and C. J. Humphries, "Measuring Biodiversity for Choosing Conservation Areas," in *Hymenoptera and Biodiversity*, ed. John LaSalle and Ian D. Gauld (Wallingford, United Kingdom: CAB International, 1993), 309–328.
18. Paul C. Mangelsdorf, *Corn: Its Origin, Evolution, and Improvement* (Cambridge, Massachusetts: Harvard University Press, 1974).

19. Paul C. Mangelsdorf, *Corn: Its Origin, Evolution, and Improvement* (Cambridge, Massachusetts: Harvard University Press, 1974), 239.

20. Paul Raeburn, *The Last Harvest* (New York: Simon & Schuster, 1995), 12.

21. E. O. Wilson, *The Diversity of Life* (Cambridge, Massachusetts: Harvard University Press, 1992), 198; William Haber, personal communication, July 1996.

22. F. Gary Stiles and Alexander F. Skutch, *A Guide to the Birds of Costa Rica* (Ithaca, New York: Cornell University Press, 1989).

23. Paul C. Mangelsdorf, *Corn: Its Origin, Evolution, and Improvement* (Cambridge, Massachusetts: Harvard University Press, 1974).

3

The Role of Values

> TVA's position would require the
> Court to balance the worth of an
> endangered species against the
> value of an ongoing public works
> measure.
>
> U. S. Supreme Court, TVA v Hill
> (1977)

The Hyannis Ponds are a group of ponds lying just north of the Village of Hyannis on Cape Cod, Massachusetts (see Figure 3-1). They include the best remaining examples in New England of the globally rare ecosystem type known as coastal plain ponds, which are relicts from the last glacier. This ecosystem has its own representative community of species, which include a number that are listed by the state as endangered, threatened, or of special concern.

During the fall semester of our course, the students enact an extensive role-play about a conservation action that Massachusetts undertook in June 1994, the eminent domain taking of 357 acres of land in and around the Hyannis Ponds complex. At the time of the taking, the land was owned by Independence Park, Inc., which had plans to develop the site for light industry. Each of the students takes on the role of a person actually involved in the events leading up to the taking. In September, we visit the Hyannis Ponds, and on the shore among the Plymouth gentians and thread-leaved sundews, the students talk to a number of people whose lives were deeply affected by the taking, some people who were in favor of the state action and some who were against it (see Figure 3-2). Later in the semester, the students read draft environmental impact reports and other documents pertaining to the situation and interview other state and local officials and private citizens involved in the situation. When the role-play itself takes place, they spend hours locked in negotiations both in and out of class.

The role-play exposes students to the values of the wide range of people involved, forcing them to examine their preconceptions about conservation. They learn that the area that includes the Hyannis Ponds is valuable not only for its biodiversity. The land around Mary Dunn Pond, the loveliest and biologically richest of the ponds, is owned by the local water company, which pumps groundwater from the area near the pond. Given the region's growing population, and the fact that some other wells have been closed due to pollution, these wells provide an important service to the local population. The developer who owns Independence Park has long stated that his goal in developing the area is to broaden the region's year-round economy and to create jobs that would allow young people to remain in the region.

Figure 3-1.
Mary Dunn Pond. This is the largest and most biologically rich of the Hyannis Ponds on Cape Cod. The plants and animals of the coastal plain ponds, taken together, constitute a globally rare community. (Photograph by Dan L. Perlman) (See also Color Plate 14)

Figure 3-2.
Many people and organizations were affected when the state of Massachusetts took the Hyannis Ponds region by eminent domain. In this photograph, Armando Carbonell (left), executive director of the Cape Cod Commission, discusses the effects of this taking with our students. (Photograph by Dan L. Perlman)

Local citizens value the land as a piece of open space, and in the past, some have used the pond shores as mountain-bike paths, a use that is extremely hard on the local flora. Part of the land that was taken, an area well away from the ponds themselves, is almost ideally located for the creation of a new exit from the main highway, and such an interchange might relieve local traffic problems. The students learn that this particular piece of property is valued by many different people for many different reasons.

THE TWO KINDS OF VALUE

We used the terms *value* and *values*, and their cognates, *valuable* and *evaluate*, many times in Chapters 1 and 2 without stopping to define them rigorously. These are terms, like *biodiversity*, *rights*, and *equality*, that are both difficult to define and often used in slippery ways by those who do not attempt a definition. In fact, the terms *value* and *values* are used in two distinct but related ways.

The first of the two meanings of *value* includes the motives, preferences, and underlying belief system that a person has in undertaking an activity, investigating a matter, or protecting an object. It is most commonly raised as an academic issue in a question such as "Is science free of values?" which means "Is science truly objective, or is it influenced by the motivations, perspectives, attitudes, or goals of the scientist?" Medicine, for example, is a science that is explicitly based on values because its goals and procedures grow from the desires of patients and the patients' subjective definitions of health. So, too, is conservation biology a science based on values; its practitioners look to the values of the communities they are trying to serve and to their own values in order to determine goals and methods necessary to achieve those goals. When commentators refer to this meaning, they typically write of *values*, in the plural.

The second meaning of *value*, the one more commonly found in the literature of economics and of biodiversity studies, concerns the worth of a particular object or activity. "What is the value of biodiversity?" is an often-asked question, and many have tried to answer it by presenting lists of the different types of value (or worth) that humans find in various elements of biodiversity. These types of value carry names such as *use value*, *option value*, and *existence value*, all terms that we discuss in this chapter. Commentators using the term *value* in this sense generally employ the singular form rather than the plural.

How important is it to distinguish between these two meanings of value in discussions about biodiversity? In order to present a clear answer to this question, it is first useful to distinguish the two concepts by separate terms. We will refer to the first kind of value with the term *values*, meaning the preferences, motivations, and belief systems that human beings use in assessing the world. We will use the term *worth* to designate the second meaning, namely, the merit and importance that humans attach to specific objects and activities.

Values, as we use the term to represent beliefs, are entirely internal to the human being, although the process of their development is molded by external events and other humans. In contrast, an individual assigns or attaches *worth* to objects and activities that are external to himself or herself. *Worth* is assigned when an object satisfies or in some way matches one of the individual's *values*. Since different people have different values, it is no surprise that they find worth in different objects and activities.

Therefore, the answer to the question, "What is the value of biodiversity?" (or in our terminology, "What is the worth of biodiversity?"), is "It depends upon the values that are important to you." An object, whether a work of art or an element of biodiversity, is not attributed great worth unless we can find a person for whom it satisfies important values.

Much of the literature we cite generally uses the term *value* to designate either or both of these concepts without distinguishing them. Although the term *worth* is not currently used in the literature, and we recognize the confusion that might arise from changes in terminology, we believe that the conceptual clarity achieved by distinguishing between values and worth is considerable. One of our *values* is clarity in thought and language, and we find that this terminology satisfies that value—and thus has great *worth*. We ask the reader to bear with us through awkward moments of translation.

The following example concerning the interactions among cleanliness, laundry detergent, and the wearing of clean shirts should help to clarify the relationship and distinction between values and worth. Cleanliness is a value that many people hold as important; it sets standards of conduct for them. If cleanliness is one of your values, then it follows that your conduct will include regular cleaning of your home, clothes, and person. Because you hold this value, the activity of wearing a clean shirt has considerable worth to you, and laundry detergent is an item of considerable worth, since it enables you to clean your shirts.

An important distinction between the two terms is that *worth* is typically conceptualized in terms of a common currency, such as dollars, while *values* almost never are—indeed, it may be impossible to place a dollar figure on one's values. To illustrate this last point, consider the question, "How much money can I pay you to care less about cleanliness?" This question, or similar ones, such as "How much money can I pay you to care less about democracy?" or "to care less about your religion?" are not questions that make any sense to us, because the underlying values, the belief system, cannot be bought. We can offer a person for whom cleanliness is an important value some money to wear a dirty shirt for a day. If we offer enough, he may accept, but we have only established a monetary figure for what he considers to be the *worth* of a clean shirt for a day; we have not put a monetary figure on his underlying values. This person may realize, upon wearing a dirty shirt for a day, that cleanliness is not as important to him as he originally thought (or that it is, indeed, more important), but this is not a pricing of values. It is a reflection of the relationship between values and experience—we accept our experiences mediated through a preexisting values system, and our values change with every important experience that impinges on them.

An object may be assigned greatly different levels of worth by different people, or even by the same person at different times, depending upon the context and particulars at hand. A clean, pressed Oxford shirt may have worth not only as a clean thing to wear but also as an item for sale by a clothier, as a bandage for an injured person, or as a multipurpose sack for a person living under subsistence conditions. On the other hand, slight changes in context and particulars may greatly alter how much worth individuals attach to an object. A man's shirt has little worth to a clothier who sells only women's clothes, a clean shirt has little worth as a bandage until a person is badly cut and cannot find a better bandage (at which point the worth increases tremendously), and the shirt's worth to the person living under subsistence conditions may be quite low as long as better sacks are available.

Consider the worth of a clean shirt to two people, one of whom holds cleanliness as an important value and one of whom does not. To the person who cares not at all for cleanliness, and is wearing a dirty shirt anyway, the worth of wearing a clean shirt is very low but could be given a price. To the person for whom cleanliness is an important value, the worth of wearing a clean shirt is much greater, and he would only give it up for a much greater amount of money. But once again, keep in mind the effect that changes in context have on evaluations of worth. Imagine going to an interview for a unique job that you have long dreamed of holding. How much money would you accept to wear a dirty shirt to that interview? Because the job itself satisfies many of your values, looking good and wearing a clean shirt hold tremendous worth in this context, even though clean shirts themselves might not be especially important for you.

Thus, values and assessments of worth are forever changing, both within groups of people and within individuals, as exemplified by the change in attitudes toward wolves and cougars in the United States over the last 400 years. These creatures have gone from having a negative worth (with bounties offered for animals killed) to having a very high worth in the eyes of many citizens, as can be seen by the increase in wolf-related tourism in Minnesota and the sums of money spent on protection and reintroduction programs for these species (see Figure 3-3). Moreover, an individual's assessment of worth may change fundamentally over time. In the short term, extreme hunger or suffering may cause an individual to completely disregard an object that would otherwise be of great worth or even consider its worth according to totally different values. Conservationist Norman Myers, when he visited our class, told our students of an ancient Arabic proverb that makes this same point: "I look at the palm tree when I am hungry and I behold the fruits. I look at the palm tree after I have eaten the fruits, and I behold that it is beautiful."

In order to analyze conservation decisions, economists have developed several systems for attaching prices to the comparative worth of activities, such as traveling to a national park as opposed to traveling to Disney World, and objects, such as the worth of the snail darter compared to the worth of building the Tellico Dam.* Some conservation decision makers have followed the lead of the economists in using such systems, the most well known of which is contingent valuation. These pricing systems, however, do not take into account the underlying values that motivate individuals to preserve the snail darter or to travel to a national park. *Values* are not amenable to such tools because they cannot be converted to a common currency that would allow for comparison. Accordingly, our survey of the biodiversity literature

---

*The question of the relative worth of the 3.5-inch-long snail darter fish (*Percina tanasi*) and the Tellico Dam was the first major test of the U.S. Endangered Species Act. The Supreme Court found for the fish in the landmark case, *Tennessee Valley Authority v. Hiram Hill, et al.*, although the dam was later completed under a direct order of Congress, destroying the only known habitat of the snail darter. It was only later that a population of the fish was discovered at another site.

A

B

uncovered only one classification that attempted to classify values and quite a few that classified the worth of various elements of biodiversity, since worth is much more amenable to comparison.

CLASSIFICATIONS OF VALUES AND WORTH

Many commentators have looked carefully at the concept of *worth* and in the process have created detailed classifications to distinguish among different types of worth. The question, "What types of worth does biodiversity possess?" has been asked and answered in many ways. We can consider the worth of any element of biodiversity, such as a species, an ecosystem, or a single organism. The California condor has worth because it is rare, because it is large, and because its line is ancient, but it is almost never thought to have worth because it can be used as medicine or because we can eat it. In contrast, the worth of the often-cited rosy periwinkle (*Catharanthus roseus*) derives from two chemicals it contains, vincristine and vinblastine, which are potent medicines against Hodgkin's disease and acute lymphocytic leukemia.[1] In addition to these species-based examples, we can think of the worth that a wetlands provides in flood control, or the symbolic worth of the oldest known living individual organism, a bristlecone pine.

In an attempt to find relationships among these different types of worth, economists have invented several larger categories, such as *use values*. As you read the next two sections, keep in mind that although these categories are described by economists as different types of "values," they lie squarely in the realm that we have described as *worth*.

Let us take a brief look at different classifications of worth that have been presented in four books that we think are representative of the many that have attempted to present such classifications: *Why Preserve Natural Variety?* by Bryan Norton, *Conserving the World's Biological Diversity* by Jeffrey A. McNeely et al., *The Economic Value of Biodiversity* by David Pearce and Dominic Moran, and *Global Biodiversity Assessment*, sponsored by the United Nations Environment Programme (UNEP).[2] In these works, different types of worth are attributed to elements of biodiversity, and these categories of worth are proposed as reasons that humans should protect biological diversity.

Alone among these commentators, Bryan Norton gives a significant amount of attention to intrinsic value, the value that an organism or species has in its own right.[3] After a lengthy discussion, he concludes that intrinsic

Figure 3-3.
Changes in the worth attributed to wolves. (A) For centuries, wolves have been depicted in popular art as vicious, brutal creatures as in this illustration of Little Red Riding Hood. (After Jean Boullet, *La Belle et la Bête* [Paris: Le Terrain Vague, 1958], illustration circa 1890 from Collection Boullet, in J. Zipes, *The Trials and Tribulations of Little Red Riding Hood* [London: Routledge, 1993]) (B) In the last couple of decades, in the United States at least, wolves have more typically been portrayed as noble or playful animals, as in this photograph from a children's book. (Dorothy Hinshaw Patent and William Muñoz, *Gray Wolf, Red Wolf* [New York: Clarion Books, 1990])

value is too problematic to use as a justification for the conservation of biodiversity. The other commentators devote their entire discussions, as Norton devotes most of his, to the different types of worth that elements of biodiversity embody for humans. Each of these types of worth provides some benefit for a being other than the element of biodiversity itself; as such, these can be considered to be types of *instrumental value*, as in Norton's terminology, or aspects of *total economic value* (the term used by both UNEP and Pearce and Moran).

Classifications of Worth

Each type of worth refers to some tangible benefit that accrues to a human being. For example the economists' *use value* is a major category that refers to the worth of items that are themselves used by humans or that contribute in some way to items that are used by humans. Economists distinguish between *direct use value* for items that are themselves consumed (e.g., timber, fish, firewood) and *indirect use value* for items that are necessary to support items that are consumed (e.g., ecosystem functions such as carbon dioxide sequestration by plants and nutrient cycling by forest microorganisms that do not themselves get consumed but that support the trees that will be harvested for lumber). *Option value*, the value of maintaining the possibility of using a resource in the future, is another type of use value.

Economists also describe a category of worth called *nonuse values* or *passive values*. This category "exists where individuals who do not intend to make use of such [biological] resources would nevertheless feel a 'loss' if they were to disappear."[4] *Existence value*, the worth that one finds in knowing that an organism or ecosystem exists even if one will never see it and gets no direct use from it, is a nonuse value. *Bequest value*, another nonuse value, is based on the knowledge that one is keeping an element of biodiversity intact for future generations.

Norton provides a case for the underappreciated importance of what he calls "transformative" value in the conservation of biodiversity. In his view, "an object has 'transformative value' . . . if it provides an occasion for examining or altering a felt preference rather than simply satisfying it."[5] In our terminology, an object can have worth in its ability to cause us to alter our values. We use Norton's book as a text in our class each year and find that the concept of transformative value resonates deeply with our students.

To help make this concept more concrete, consider the examples of the different types of worth that one might attribute to a mature redwood tree as depicted in Figure 3-4.

A Classification of Values

In our review of the literature, we have uncovered only one classification of what we are calling values. Stephen R. Kellert provides this classification of values in *The Value of Life* and elsewhere.[6] Kellert's use of the term *values* is quite close to our own. In discussing an early study that he did for the U.S. Fish and Wildlife Service, he describes "developing a means for classifying and measuring people's values of wildlife and nature,"[7] which led to his appreciation that humans have "a set of basic values toward animals and the natural environment."[8] (See Table 3-1 for a list of the nine values he recog-

Figure 3-4.
Sample statements of different types of worth that might be attributed to a large redwood tree.

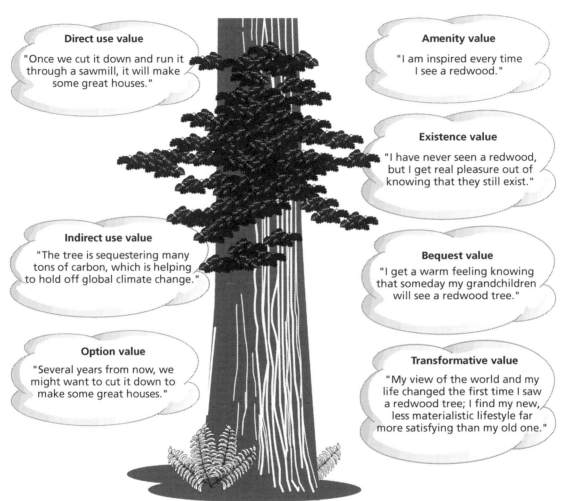

Direct use value

"Once we cut it down and run it through a sawmill, it will make some great houses."

Amenity value

"I am inspired every time I see a redwood."

Existence value

"I have never seen a redwood, but I get real pleasure out of knowing that they still exist."

Indirect use value

"The tree is sequestering many tons of carbon, which is helping to hold off global climate change."

Bequest value

"I get a warm feeling knowing that someday my grandchildren will see a redwood tree."

Option value

"Several years from now, we might want to cut it down to make some great houses."

Transformative value

"My view of the world and my life changed the first time I saw a redwood tree; I find my new, less materialistic lifestyle far more satisfying than my old one."

nizes.) Kellert claims that the values are "rooted in human biology," as well as "shaped by the formative influence of experience, learning, and culture."[9] The list includes both positive and negative values toward nature.[10] Kellert clearly sees the values he describes as being located in the psyches of human beings, rather than being attached to external objects (as is the *worth* of the other classifications discussed). His classification of values describes the types of conduct that individuals undertake, and the emotions that they feel, toward nature.

Just as the classifications of worth imply that different actors can attribute different kinds of worth to a single object, Kellert's list of values enables several actors to view a single object from radically different perspectives. By starting with different values, each actor arrives at a different assessment of the worth of the object. Even when an object is highly valued by several individuals, each may employ a different perspective to reach his or her conclusion, and each may focus on different features of the object in question. In fact, depending on

Table 3-1.   Kellert's classification of human values of nature.

Type of Value	Definition
Utilitarian	Practical and material exploitation of nature
Naturalistic	Satisfaction from direct experience/contact with nature
Ecologistic-Scientific	Systematic study of structure, function, and relationship in nature
Aesthetic	Physical appeal and beauty of nature
Symbolic	Use of nature for metaphorical expression, language, expressive thought
Humanistic	Strong affection, emotional attachment, "love" for nature
Moralistic	Strong affinity, spiritual reverence, ethical concern for nature
Dominionistic	Mastery, physical control, dominance of nature
Negativistic	Fear, aversion, alienation from nature

SOURCE: Abstracted from Stephen R. Kellert, "The Biological Basis for Human Values of Nature," in *The Biophilia Hypothesis*, ed. Stephen R. Kellert and Edward O. Wilson (Washington, D.C.: Island Press, 1993), Table 2.1, p. 59.

Figure 3-5.
Different perceptions of deer based on the values held by different individuals; the values are taken from Kellert's classification.

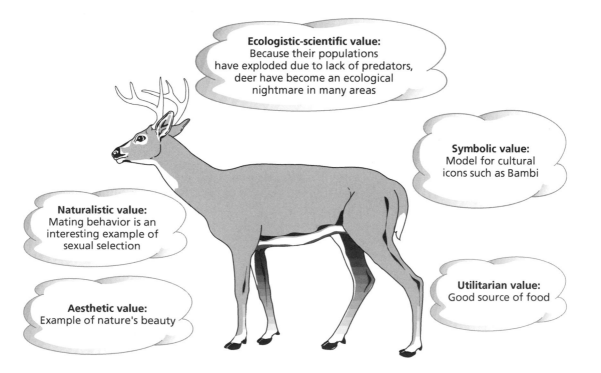

the observers and their values, a single item may hold great positive worth to some and great negative worth to others. Once again, we see that the worth attributed to an object is an expression of the values that an individual holds. For example, deer in North America can be evaluated in several different ways, depending on the individual's values, as shown in Figure 3-5.

Our purpose in this section was to present a brief overview of several attempts to classify categories of worth and values, especially as they pertain to biodiversity. These attempts are important because conservation decision makers need to articulate the values underlying their decisions and the types of worth they seek to protect through their efforts. The following accounts of the search for a rare plant and changes in the perception of a previously unimportant tree species illustrate the importance of values and worth in conservation decision making.

CONFLICTS AMONG VALUES AND CHANGES IN WORTH: TWO BOTANICAL TALES

To clarify the kinds of issues that can arise when decision makers hold different values and attach different types of worth to elements of biodiversity, we present the following two tales. In the first, we follow a botanical expedition of our colleague Michael Donoghue, a botany professor at Harvard University and Director of the Harvard University Herbaria.[11] Donoghue has spent the bulk of his professional career analyzing the characteristics of plants in order to understand the phylogenetic (or evolutionary) relationships among different plant taxa. These characteristics have traditionally included aspects of plant morphology, anatomy, and development. The primary starting point of phylogenetic analysis for botanists has been the herbarium sheet, which consists of a dried and pressed plant specimen mounted on heavy, acid-free paper that is stored in a herbarium. However, within the last 10 years, the powerful tool of molecular analysis, which generally requires a sample of leafy tissue taken from a live plant to allow extraction of proteins or DNA, has been added to the phylogeneticist's repertoire.

Donoghue has concentrated his systematic work on the family Caprifoliaceae. He has demonstrated that certain genera traditionally placed in the Caprifoliaceae do not belong in this family; rather they belong in the family Adoxaceae, which also contains two little-known genera of a single species each: *Tetradoxa* and *Sinadoxa* (see Figure 3-6). Both *Tetradoxa* and *Sinadoxa* are known only from a very few collected specimens, and both contain important information for our understanding of the evolution of the family Adoxaceae, and perhaps the families most closely related to it. The only known locations for plants of these two genera are in the mountainous regions of the interior of China. In the summer of 1995, Donoghue and several other botanists set out for the interior of China in hopes of observing *Tetradoxa omeinsis* in its natural habitat on Mt. Omei in the province of Szechuan. They intended to gather specimens for several herbaria and leafy material for molecular analysis. The group, which included several Chinese botanists, began its search for *T. omeinsis* with just one

Figure 3-6.

A cladogram of Adoxaceae, showing the position of *Sinadoxa* and *Tetradoxa*. (Adapted from personal communication with Michael Donoghue, 1996.)

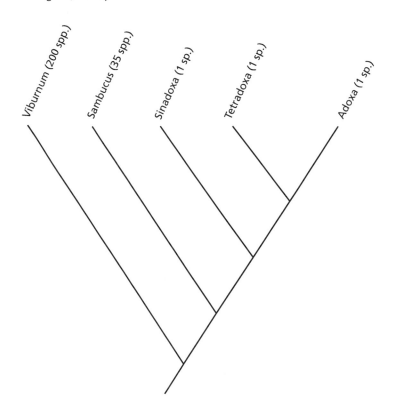

clue to its whereabouts—the location information from the single collection that represented the sole instance of a human being knowingly observing this species.

After searching for several days in this small area, the team was able to find only two patches of *T. omeinsis*, one on each of two large rocks near a heavily traveled trail that ran between two Buddhist temples. The party of botanists concluded that these were the only two patches of the species *Tetradoxa omeinsis* they were going to find, which presented the distinct possibility that these were the only two examples of this species on Earth.* The group was thus faced with a difficult decision: Should they make a collection, and if so, how much of a collection? To better understand the decision-making process that the botanists underwent, we should consider the underlying values that helped guide their actions.

---

*At the October 1996 World Conservation Union (IUCN) meeting, the national government of the People's Republic of China nominated Mt. Omei for designation as a World Heritage site, both because of the presence of *Tetradoxa omeinsis* and because of the importance of its Buddhist temples. (Ben Tan, personal communication 1996.)

Botanists from the United States and botanists from China share many of the same values in their work—both groups wish to understand the evolutionary processes that have been at work on the plants in their region and in the world, both are interested in the ecological functioning of plants in their environment, and both are interested in understanding the various services that plants can provide to society through foods, medicines, and structural materials. However, cultures may rank the relative importance of these values differently. For example, China has an ancient tradition of the use of plants in medicine that continues full strength to this day and deeply motivates modern Chinese botanists. Although there had been a similar tradition in the West, that tradition has substantially dissipated over the last 200 years. Accordingly, although one of the reasons currently put forth in the West for the conservation of biodiversity is the possibility of discovering new drugs from little-known species, Western botanists do not have the same direct concern as do Chinese botanists to immediately evaluate the medicinal uses of each newly discovered plant.

Two of the Chinese botanists were very interested in the use of plants in Chinese medicine, a form of Kellert's utilitarian value that might be called "helping humans through medicinal use of plants." The decision on how much to collect was driven by a rich combination of values, including some that might be called "the advancement of human knowledge through science" (part of Kellert's ecologistic-scientific value), "the preservation of species in their natural habitats" (perhaps a combination of Kellert's naturalistic and moralistic values), and "the respect for cross-cultural values" (not a part of Kellert's classification, but an important value nonetheless). These values set the standards of conduct that scientists in such a group expect themselves to satisfy with regard to other people, relationships, and things. But in this particular instance these values came into conflict. Each of the botanists held values that led him to want to take several specimens of the plants. Yet each also valued the preservation of the species, and recognized that this value would best be served by making a minimal collection or no collection whatsoever.

Despite the benefits that multiple herbarium sheets would have for the scientific community, and despite the risk inherent in collecting so small a sample size for molecular analysis, the group only collected two of the 15 flowering shoots from the larger of the two patches, enough material for just two herbarium sheets, and plucked and preserved two additional leaves for later molecular analysis. In this case, given the fragile nature of the only known population of the plant, collections for medicinal use were not made.

Based on the values that the various botanists held, we can describe several different types of worth that might be ascribed to *T. omeinsis*. For example, there is the unknown worth that the plant could have for medicinal purposes. In the Western world, this worth would be relegated to the rather unlikely possibility that the plant contains a secondary chemical compound that could be used to safely treat a disease recognized by Western doctors and that the drug derived from the chemical could pass rigorous clinical trials. However, in China, where a large number of plants and animals find a use in traditional medical practice, the likelihood of medicinal use is much higher (although this worth must be discounted by the fact that with only two patches of the plant known, the species could easily be extirpated during its first days of medical use). There is a second worth, an indirect use value, that *T. omeinsis* might have in its native habitat—it may be important in some way either to the functioning of its ecosystem or to the continued existence of another species that is of direct use to humans. Removal of these plants from their habitat could affect some of the remaining plants and animals in that habitat, such as a species of flying insect that pollinates them. However, it is unlikely that the loss of a plant species as rare as *T. omeinsis* would significantly alter the functioning of the ecosystem or cause another species to go extinct. A third important worth is existence value, the worth a species derives from human knowledge of its continued existence, even when the species in question has no other demonstrated worth. How much existence value *T. omeinsis* possesses is difficult to assess, since only several dozen botanists know of its existence. However, for these people, the existence of the species is quite important. Perhaps this worth might best be considered a type of nonconsumptive use value, since *T. omeinsis* has tremendous worth to plant systematists and other evolutionary biologists because of what it can tell us about the evolutionary history of the Adoxaceae and its related families.

We chose this example because the literature tends to stress worth in justifying the conservation of biodiversity. In the story of *Tetradoxa omeinsis*, however, values play the primary role. They are the motivating force for the decisions, because they are deeply held. Accordingly, values are less likely to change with changes in context than are assessments of worth. The types of worth mentioned in the preceding paragraph are much more tenuous and context dependent than the underlying values. If an important medicine were discovered from a specimen of *T. omeinsis*, the worth of the two known plants would increase drastically; if several populations containing thousands of individuals were discovered on other mountains, the worth of these two plants would plummet. In fact, just such a dramatic change in worth of a species of plant occurred in North America in the late 1980s. The story of the Pacific yew demonstrates how evaluations of worth can fluctuate in response to changes in context.

For tens of thousands of years, Pacific yew trees (*Taxus taxifolia*) functioned as a part of their ecosystem, the old-growth coniferous forest of the Pacific Northwest. The species was what foresters would call an associate member of this ecosystem, not as numerous in individual trees nor as important in biomass as the dominant elements of this ecosystem, such as the Douglas-fir (*Pseudotsuga menziesii*), the red cedar (*Thuja plicata*), the Sitka spruce (*Picea sitchensis*), or any of a dozen other of the imposing conifers of the Pacific Northwest. Exactly what the Pacific yew's importance to the ecosystem was, and is, evades easy assessment. It may have a crucial ecological role to play that could not be taken up by any other species, or it may be largely redundant, in the sense that had it been extirpated naturally, another species or suite of species would have expanded its niche and relative population size in order to replace whatever functions the yew has in the ecosystem. For the last 130 years, the Pacific yew had a specific worth to a certain band of humans who inhabited the forests of the Pacific Northwest. That is, to the lumberjacks and foresters the yew was a trash species, hardly worth bothering with in their attempts to cut the highly prized mighty Douglas-firs and cedars. When the trunks of the Pacific yew were not burned or left to rot, they were used for pulp or matchsticks. For all these years up until the late 1980s, the Pacific yew had a more or less verifiable worth that could be measured in the market, and this worth was quite low. In addition, during that time, it maintained some worth due to its role in the functioning of its ecosystem, as well as the worth it had to Native Americans and to conservationists as an element of biodiversity growing in the wilderness.

In the late 1980s, Western science discovered that the chemical taxol was an important treatment agent in the fight against ovarian cancer. Taxol was only known to be present in commercially harvestable concentrations in the bark of the Pacific yew, and the bark of up to 12 mature yews more than 100 years old was needed to treat a single patient.[12] Virtually overnight, a species that had been disregarded in the economic marketplace for a century came to have such enormous worth that, had the market been left to proceed untrammeled, the species would likely have been harvested to extinction in a few years. Because our society values human life, once scientists discovered the role that the Pacific yew can play in prolonging the lives of women suffering from cancer, the worth accorded to the species grew enormously.

In this chapter we drew a distinction between *values*, which are held by individual humans, and *worth*, which is attributed to objects in the world by humans. Throughout the rest of the book, we explore the role that values and related concepts play in creating biodiversity inventories and setting priorities for the protection of biodiversity. In the next chapter, we discuss the role that values play in the ways in which people use, and define, the term *diversity*.

References

1. Edward O. Wilson, *The Diversity of Life* (Cambridge, Massachusetts: Harvard University Press, 1992), 283.
2. Bryan G. Norton, *Why Preserve Natural Variety?* (Princeton, New Jersey: Princeton University Press, 1987); Jeffrey A. McNeely, Kenton R. Miller, Walter V. Reid, Russell A. Mittermeier, and Timothy B. Werner, *Conserving the World's Biological Diversity* (Gland, Switzerland: IUCN, 1990); David Pearce and Dominic Moran, *The Economic Value of Biodiversity* (London: Earthscan Publications, 1994); United Nations Environment Programme, *Global Biodiversity Assessment* (Cambridge: Cambridge University Press, 1995).
3. Bryan G. Norton, *Why Preserve Natural Variety?* (Princeton, New Jersey: Princeton University Press, 1987), 135–182.
4. United Nations Environment Programme, *Global Biodiversity Assessment* (Cambridge: Cambridge University Press, 1995), 830.
5. Bryan G. Norton, *Why Preserve Natural Variety?* (Princeton, New Jersey: Princeton University Press, 1987), 10.
6. Stephen R. Kellert, *The Value of Life* (Washington, D.C.: Island Press, 1995); Stephen R. Kellert, "The Biological Basis for Human Values of Nature," in *The Biophilia Hypothesis*, ed. Stephen R. Kellert and Edward O. Wilson (Washington, D.C.: Island Press, 1993), 43–44.
6. Stephen R. Kellert, "The Biological Basis for Human Values of Nature," in *The Biophilia Hypothesis*, ed. Stephen R. Kellert and Edward O. Wilson (Washington, D.C.: Island Press, 1993).
7. Stephen R. Kellert, *The Value of Life* (Washington, D.C.: Island Press, 1995), 5.
8. Stephen R. Kellert, *The Value of Life* (Washington, D.C.: Island Press, 1995), 6.
9. Stephen R. Kellert, *The Value of Life* (Washington, D.C.: Island Press, 1995), 9.
10. Stephen R. Kellert, *The Value of Life* (Washington, D.C.: Island Press, 1995), Chapter 2.
11. The details of this account come from a public seminar that Michael Donoghue gave, as well as from personal communications that we have had with him throughout 1996.
12. Christopher Joyce, "Taxol: Search for a Cancer Drug," *BioScience* 43 (1993): 133.

4

Diversity

> "Diversity" is a term with no
> essential philosophical, political,
> or aesthetic content.
>
> —Louis Menand, "Diversity"[1]

SEARCHING FOR DIVERSITY

The most difficult day of each academic year is the third Monday in September, when we have to pare an applicant pool generally consisting of 40 gifted and highly motivated students down to the 16 who will be members of our class until the next May. In order to assure quality instruction, especially out in the field, we need to keep our class size manageable. As we wade through the emotionally draining process of ranking and reranking students, we keep in mind that it is a *class* we are selecting, not just an independent assortment of qualified students, and one of the qualities that we actively seek in creating a class is *diversity*. Although the course is offered jointly by two departments, Environmental Science and Public Policy (ESPP) and Biology, we welcome applications from students across the college. We have discovered the importance of including a diversity of perspectives and backgrounds from students in a range of disciplines such as chemistry, economics, anthropology, religion, archaeology, art, mathematics, philosophy, and engineering. But we consider more than academic affiliations in our attempts to create a diverse class. We pay attention to several aspects of the applicants' backgrounds: the settings in which they grew up (urban, suburban, rural), their nationality and the culture in which they were raised, their previous coursework, their gender, and their year in school. Across all of these areas, we look to assemble a diverse class, even if it means that we have to turn away top-notch candidates. We may, for example, take an enthusiastic sophomore with little background instead of a well-trained, and apparently better-qualified, senior—largely because we generally have fewer sophomores and know from experience that diversity of ages often is a positive addition to the class dynamic.

Although we value diversity, we do not seek a class that is diverse in all ways. Regarding some features of our applicants, we intentionally seek to *decrease* diversity. We aim for a class that shows little diversity in features such as interest in the subject, commitment to the physical and intellectual rigors of the course, writing ability (or the deep desire to improve one's writing), and ability to get along with others. We do not want diversity in our students' commitment to the course and interest in the subject matter; we want *all* of our students to take the course very seriously. And finally, there are the vast majority of human features that we simply do not notice; an applicant's blood type, hair color, size, handedness (left or right), and musical ability play no role in our selection process. We appreciate students with strong backs for carrying gear on our field trips, and we took great pleasure in hearing one of our students give a violin recital one year in Costa Rica, but we do not consider these attributes in September when we try to put together a diverse class (see Figure 4-1).

The larger universe in which our class exists—namely, an undergraduate college—also strives to create a diverse incoming class each year, but the college's criteria are somewhat different from ours. Size, strength, and musical ability are important attributes that receive considerable attention in the application process; athletic coaches, conductors, and alumni would be displeased with an incoming class that exhibited reduced diversity in these areas—especially if the missing elements of diversity were among the tallest, most muscular, or most musically talented applicants. On the other hand, most colleges do not aim for diversity in SAT scores—they want their students clustered toward the high end of the scale.

What, then, do we strive for when we select our class? Are we looking for diversity per se? No, we seek to create a diverse class, but we do not want one that is diverse in all regards. We want one that is diverse in only a few features, features that we value in a particular way, the diversity of which will increase the educational worth of the class to all students. An economics major, for example, brings to the class a world of information and modes of thinking that most of the other students only know in vague

Figure 4-1.
Three ways of looking at diversity. Although *diversity* is often used as a term with only positive connotations, there are many circumstances in which diversity can be thought of as either neutral or negative. An observer in a specific context may react in one of three different ways toward the diversity of any chosen feature: she may place a positive value on it; she may ignore it; or she may place a negative value on it, that is, she may seek as little diversity as possible in that feature. This figure depicts the features of students who apply to our class each year and how we categorize the diversity of those features.

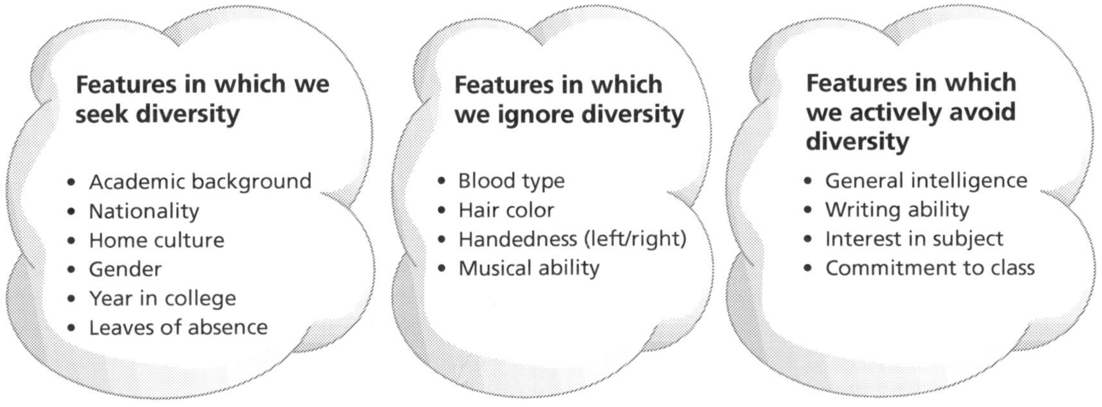

Features in which we seek diversity

- Academic background
- Nationality
- Home culture
- Gender
- Year in college
- Leaves of absence

Features in which we ignore diversity

- Blood type
- Hair color
- Handedness (left/right)
- Musical ability

Features in which we actively avoid diversity

- General intelligence
- Writing ability
- Interest in subject
- Commitment to class

outline. This information and these ways of thinking can enrich our discussions of tropical deforestation, intellectual property rights, and reserve design. Interestingly, even in some of the features where we aim for diversity, as with academic major, we do not attempt to maximize diversity above all—if that were the case, our class of 16 students would comprise one student from each of 16 academic majors; in practice, however, we usually have a disproportionate number from our sponsoring departments of ESPP and Biology. We weigh the worth of having many diverse views and backgrounds against the worth of having a class in which a large percentage of the students have strong academic backgrounds in biology and environmental policy.

THE FOREST AND THE TREES

All we said in the preceding paragraph about student diversity is true for biodiversity. There are a tremendous number of features in which the organisms of the world exhibit diversity, but conservation decision makers do not care equally about all of them. And for some features of organisms or ecosystems or other biological elements, conservation decision makers will actively seek to *minimize* diversity.

Consider the trees in a forest; conservationists pay attention to certain ways in which any collection of trees in a forest are diverse, and ignore other ways in which they are diverse. They consider the genetic diversity of the forest valuable, although its details are almost always unknown. They generally consider a diversity of age groups among the trees (i.e., seed bank, seedling, sapling, understory, overstory, and overmature) to be more valuable than a single-aged stand, if for no other reason than it is less likely that all the trees in the stand will die off at once. Conservationists have always considered a diversity of species to be valuable. At a different level of analysis, a greater diversity of higher taxonomic groups, such as genera and families, is more valuable than a lesser diversity.

There are other characteristics in which the trees exhibit diversity, but to which no one pays any attention. A tree randomly chosen from a forest will possess a certain number of leaves, a number that is in all probability different from the number of leaves on any other tree. A leaf randomly chosen from that tree will contain a certain number of cells, and there will be a verifiable diversity in the number of cells per leaf on that particular tree; in addition, each leaf cell contains a certain number of chloroplasts. In any particular forest, there is presumably great diversity in these features, although we cannot be sure, because, as no one values this diversity, no one has rigorously measured the diversity of a forest considering these features (see Figure 4-2).

Finally, there are a few features of forest trees, such as their health, in which conservationists value a clustered (i.e., nondiverse) distribution; they do not value all possible states of health equally. "Health" is a complex and subjective concept, so for the sake of simplicity we will define an organism to be healthy if its ability to survive and reproduce has not been seriously compromised by disease or physical damage, and to be unhealthy if such compromise has

Figure 4-2.
A forest with a diversity of leaf, cell, and chloroplast numbers. A forest with a high diversity of species in it is valued by both ecologists and conservationists. But diversity in other features of the forest has been ignored by ecologists and conservationists. Some forests have higher diversity in the feature "number of leaves per tree" than other forests, but no one uses that fact to set conservation priorities. There are countless ways in which a forest is diverse—for example, number of cells per leaf, number of chloroplasts per cell—that are unvalued in conservation.

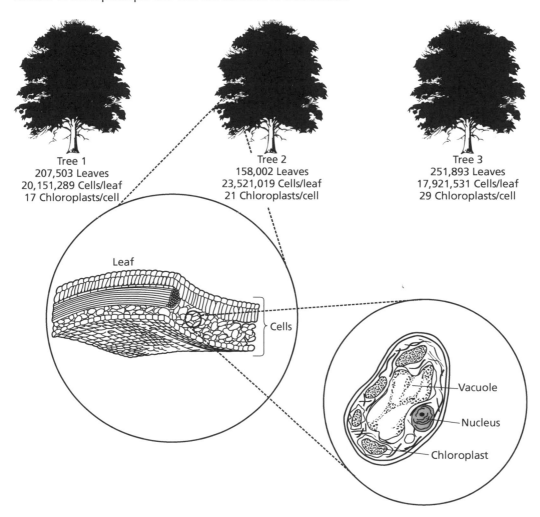

Tree 1
207,503 Leaves
20,151,289 Cells/leaf
17 Chloroplasts/cell

Tree 2
158,002 Leaves
23,521,019 Cells/leaf
21 Chloroplasts/cell

Tree 3
251,893 Leaves
17,921,531 Cells/leaf
29 Chloroplasts/cell

Leaf

Cells

Vacuole

Nucleus

Chloroplast

occurred. As an example, one of the most endangered tree species in eastern North America is *Torreya taxifolia*, the stinking yew, which persists only in a few populations on the Florida panhandle and southern Georgia. Up until the early 1960s, most of the individual trees of this species were healthy—there was not a high diversity in healthiness. Unfortunately, in the early 1960s, a fungus that caused stem and needle blight struck the remaining populations and to this day seriously compromises the continued existence of the species. The diversity of healthiness increased—that is, the populations went from having mostly healthy trees, a low level of diversity in health, to having a more equal number of healthy and unhealthy trees, a higher level of diversity in health—but not to the gratification of any conservationist.[2]

THE INTRINSIC RELATIONSHIP BETWEEN DIVERSITY AND VALUES

We devoted Chapter 3 to a discussion of values and how they affect conservation decision making. The purpose of the current chapter is to extend that argument to the concept of diversity itself. Diversity exists among a nearly infinite number of features; but as a concept approaches infinite applicability, it also approaches meaninglessness. This near infinity needs to be broken down in the following manner. In our attempts to conserve a diverse assemblage of elements of the natural world, there are many features that we pay attention to and even more that we ignore. Of the features that we pay attention to, there are many whose diversity we value and others whose diversity we actively disvalue. The choice of which features to pay attention to and which diversities to value is one that derives from the values and goals of individuals.

Even among "obvious" cases of difference, investigators may choose to ignore diversity in a feature. As ecologist F. W. Preston says of ecological diversity, which he equates with species richness, "Two individuals are not diverse if they are of different sexes, but only if they are of different species. Similarly, subspecies do not count as diverse, nor are larvae from adults."[3] In this context, Preston has made an explicit judgment about the features of diversity that he values and considers worthy of attention and those that he disregards; he explicitly ignores, for his purposes, obvious examples of differences such as sex and life stages within a species. According to his system of values, the presence of seedlings and saplings, mentioned earlier as a valuable diversity to many, would confer no added diversity to a forest with only mature trees. In contrast, for the conservationist interested in functional diversity, juveniles and adults of both insects and trees often inhabit entirely different ecological niches, thereby contributing significantly to diversity (see Figure 4-3).

In less obvious cases, as with the number of leaves on a tree or the cells in a leaf, diversity is almost always ignored. The leaves on a tree are considered a nondiverse assemblage. As we have shown earlier, diversity is there—but no one values it. We devote the rest of this chapter to a set of cases in which the valuing of diversity might be problematic, even among conservation decision makers. This set is meant to be illustrative and not exhaustive. We first look at ecosystem integrity, which we contend is a feature of the natural world the diversity of which conservation decision makers do not value. We then look at the current decrease in the varieties of apples and suggest it as an example of features that most conservationists would ignore, but whose diversity some conservationists would value. Next we take a close look at the concept of species richness; there we explore how the different values of conservationists and community ecologists have become confused around the term *diversity*, to the detriment of conservation. In our final example, we speculate on whether the reforestation of farmland in Vermont adds to or detracts from biodiversity.

Figure 4-3.
Standard measures do not capture all diversity. Individuals of many species change fundamentally during their development, as can be seen by comparing (A) the yellow birch (*Betula alleghaniensis*) seedling with (B) the 200+ year old adult of the same species. Similarly, individuals of different castes in social insect colonies, such as the two leaf cutter ant workers in (C) can play very different roles, and add to diversity of forms without being accounted for in standard measures. (See Color Plates 4–7 for further examples; Photographs by Dan L. Perlman)

A

C

B

ECOSYSTEM INTEGRITY

The Nature Conservancy, the largest private land protection organization in the United States, has stated that its mission is to save "the last of the least and the best of the rest." The last of the least is synonymous with "most endangered." The best of the rest, as demonstrated through the Nature Conservancy's own practices, means the finest examples of a region's representative ecosystems, or to use the term most commonly applied, those examples with the greatest "ecosystem integrity."

Integrity in the biological sense, according to Karr and Dudley, refers to

a system's wholeness, including presence of all appropriate elements and occurrence of all processes at appropriate rates. Whereas diversity is a collective property of system elements, integrity is a synthetic property of the system.[4]

The Hyannis Ponds, which we introduced at the start of Chapter 3, are considered the coastal plain pond ecosystem in New England with the highest level of ecosystem integrity. There are many other coastal plain ponds in New England, but most of them are in various stages of degradation, that is, they have low levels of ecosystem integrity, mostly from human recreational and industrial use, pollution, and the invasion of introduced species (see Figure 4-4).

Neither the Nature Conservancy nor any other conservation organization suggests that we try to conserve a diverse range in integrity of particular ecosystem types. That is, no one suggests conserving, in addition to the high-integrity example of the Hyannis Ponds, an example with moderately high integrity, an example with moderately low integrity, and a totally degraded example. We do not proceed according to this strategy, not because the latter three examples do not truly add to biodiversity—after all, they are all biological systems—but because we value ecosystems with low integrity less. This is because, among other reasons, they do not function well; they lack many of the interesting species that are characteristic of this ecosystem type, species that cannot be found elsewhere; and to many people, they are unattractive. Although these reasons and the conclusion they warrant may seem self-evident to most conservation decision makers, they are nonetheless based on a set of values. There is no value-free "diversity principle" that tells us not to value diversity in ecosystem integrity and to value diversity of species—our values and goals tell us.

APPLE VARIETIES

In 1910, nurseries in the United States stocked approximately 500 varieties of apples, and 35 of those varieties made up 90% of the apple crop. Varieties back then included, in addition to the Macintosh, Northern Spy, and Jonathan that we still know today, such unremembered names as Crow Egg, Rambo, Roxbury, and Wolf River.[5] Today, there are only 100 apple varieties sold by commercial nurseries, and one variety that did not even exist in 1910, the Delicious, makes up 25% of the crop.[6] Modern methods of harvesting, storing,

Figure 4-4.
Location and integrity of coastal plain pond shore communities in southeastern Massachusetts. The Hyannis Ponds (largest black dot) have the highest ecosystem integrity of the remaining coastal plain ponds in their area. (Map used with permission of the New England Chapter of The Nature Conservancy)

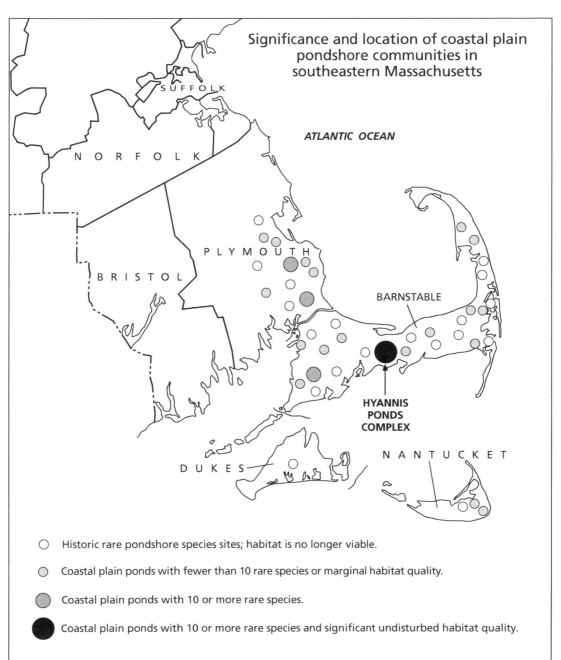

Significance and location of coastal plain pondshore communities in southeastern Massachusetts

○ Historic rare pondshore species sites; habitat is no longer viable.

◉ Coastal plain ponds with fewer than 10 rare species or marginal habitat quality.

◉ Coastal plain ponds with 10 or more rare species.

● Coastal plain ponds with 10 or more rare species and significant undisturbed habitat quality.

and marketing have made it financially impractical for apple growers to produce any but the 10 to 12 most common varieties (see Figure 4-5).

Biodiversity is decreasing—in terms of apple varieties—but not many conservation decision makers care. For most conservation organizations, these are biological features to ignore, and the diversity present within them is not considered to be part of the biodiversity that they set out to protect. This attitude is perfectly consonant with the values of these organizations. Human-produced biological diversity can be considered less valuable than natural biodiversity for a number of reasons: because humans will often have the capabilities of recreating biodiversity that they created in the first place, because natural biodiversity is more spiritually inspiring, and because natural biodiversity is part of an autonomously functioning system.

However, some organizations, such as the Worcester Horticultural Society of Massachusetts, have begun to dedicate orchards to preserving apple varieties that are in serious decline, just as other organizations are working to save particular cultivars of roses, breeds of dogs, and breeds of livestock that are disappearing. No one can question that diversity exists in each of these cases, and that the diversity is biological. This is biodiversity that is unvalued, rather than disvalued—for although few conservationists would actively wish to see diversity of breeds and cultivars decrease, which makes this a different case from the preceding ecosystem integrity example, few would care to devote any resources or efforts to their preservation.

A more complex case of how different values can alter the meaning of diversity is that of the difference between a community ecologist's view of diversity, often expressed as species richness, and the conservationist's view of biodiversity. First we look closely at species richness itself, in order to show that it is not as objective at it seems, and then we examine the differences between the two terms, and the confusion that sometimes occurs in their usage.

SPECIES RICHNESS

Think back to the example of the impatiens in the Monteverde Cloud Forest Preserve in Chapter 2. One way of assessing the biodiversity of that preserve

Figure 4-5.
Apple varieties. A selection of the varieties being preserved at the orchards of the Worcester Horticultural Society: Crow Egg, Rambo, Roxbury, and Wolf River (clockwise from small apple at top). (Photograph by Glenn Adelson)

is simply to count the number of species in it. Such a species count, known technically as species richness, is the most commonly used measure of biodiversity today. One might argue that species richness is free of biases and values—objectively, there are *x* and only *x* species in a given area. Even this simple assessment, however, has implicit in it a number of very powerful biases, decisions, and value statements that should be, but almost never are, made explicit. In choosing to count species in a region, an investigator decides:

1. to count only certain groups of species (e.g., birds or trees) and not to count other types of species (e.g., bacteria and fungi);

2. to give all species equal weight in the measure, so that no species counts for more than any other;

3. to employ a particular species concept to distinguish species (see Chapter 6, pages 118–122, for more on species concepts):

4. *not* to count other levels of biological diversity (e.g., subspecies, taxonomic orders, genes, or ecological communities);

5. where the boundaries of the region are, thereby disregarding neighboring areas and the additional species they contain;

6. what times of year to count, or over how many years to count.

Each of these decisions has an impact on the final species count, and each is shaped by the investigator's values, interests, biases, or resources (see Figure 4-6).* Slight changes in any of these value-laden decisions alter the result of the count, and decision makers might attribute considerably more (or less) worth to the region if the count changed. The tool of species richness has been used to answer particular questions in the discipline of community ecology, the study of the mechanisms that regulate change in ecosystems or change in species diversity within ecosystems. However, conservation biology has picked up species richness as its commonest tool for measuring biodiversity. Robert May explains its attractiveness, and hints at its danger, in his chapter of the edited volume *Biodiversity: Measurement and Estimation:*

> Biological diversity can be quantified . . . at many different levels. Commonly . . . we choose the numbers of species. This is sensible. . . . Effective action needs money and money ultimately depends on widespread support among the general public. It is easier to recognize the biodiversity immanent in species—especially charismatic vertebrates or colorful plants—than in gene pools or ecosystems. . . . The basic message . . . is that biological diversity has many dimensions. Summarizing it by a simple species count, *as is done in the rest of my chapter*, can often obscure conceptual understanding, and can sometimes do harm in practice. [Emphasis added][7]

We would emphasize May's concerns more than his method. When using a tool, it is important to understand the toolmaker's goals and to be wary

---

*See also Chapter 4, "Ecological Science Is Value Laden," in K. S. Shrader-Frechette and E. D. McCoy, *Method in Ecology* (Cambridge: Cambridge University Press, 1993).

Figure 4-6.

The diversity typically valued and ignored by studies relying on species richness. Such studies claim to be measuring "diversity." However, they actually measure only a tiny subset of diversity, and in fact only a small part of species diversity.

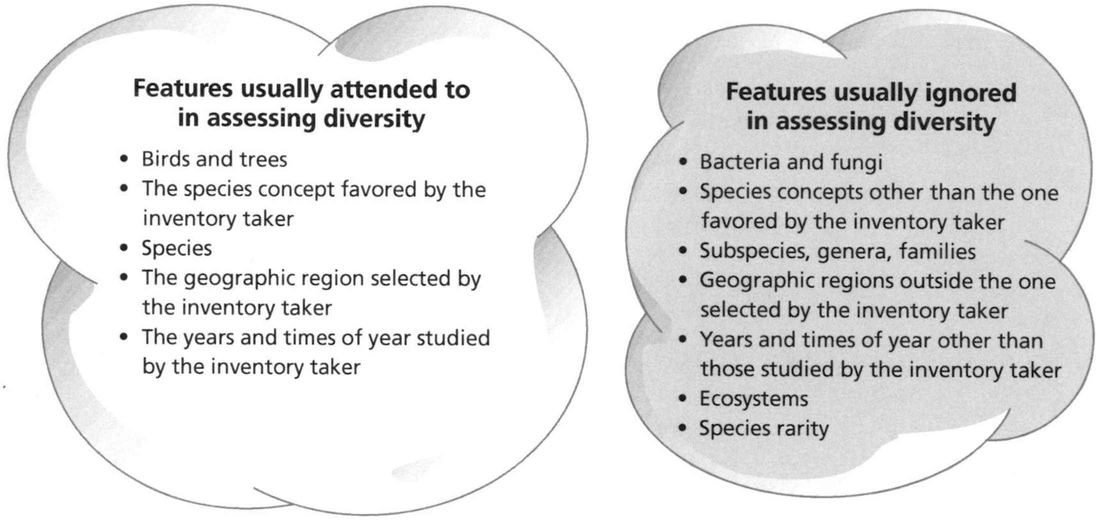

Features usually attended to in assessing diversity

- Birds and trees
- The species concept favored by the inventory taker
- Species
- The geographic region selected by the inventory taker
- The years and times of year studied by the inventory taker

Features usually ignored in assessing diversity

- Bacteria and fungi
- Species concepts other than the one favored by the inventory taker
- Subspecies, genera, families
- Geographic regions outside the one selected by the inventory taker
- Years and times of year other than those studied by the inventory taker
- Ecosystems
- Species rarity

when using the tool for purposes other than those for which it was made. Accordingly, a look at the differences between the values and goals of community ecologists and those of conservationists can help us to see why the concept of species richness is of only limited use to the conservation decision maker.

THE COMMUNITY ECOLOGIST'S *DIVERSITY* AND THE CONSERVATIONIST'S *BIODIVERSITY*

The discipline of community ecology has a long history of field studies and theoretical discussion of *diversity*, most of which preceded the coining of the term *biodiversity*. Here we briefly review some relevant aspects of the concept *diversity* to illustrate how these differ from the ways in which *biodiversity* is defined and assessed by conservationists.

The following, from D. Walker's paper, "Diversity and Stability," is a typical definition of diversity in terms of species richness. Note that equitability, a comparison of the number of individuals of each species at a site, is often included in definitions of diversity:

> The *diversity* of an assemblage is usually broken down into the number of species in it (species richness), and the distribution of the number of individuals amongst the species (equitability); for the most part I shall be concerned with the former, to which I shall restrict the term diversity.[8]

Compare this with any of the definitions of biodiversity presented in Chapter 1, such as this one from the International Council for Bird Preservation:

> Biodiversity is the total variety of life on earth. It includes all genes, species and ecosystems and the ecological processes of which they are part.[9]

Two differences between the terms *diversity* (as employed by community ecologists) and *biodiversity* strike us as most telling: 1) the manner in which individual elements of diversity and biodiversity, especially species, are treated by ecologists and conservationists, and 2) the reasons stated for being interested in diversity or biodiversity. In the first matter, the diversity measures of ecologists treat all species as equivalent. In fact, as Walker states, only two types of information are used as input for such measures: the number of species, and the equitability of distribution of numbers of individuals among these species. The identity of the species in question is of no concern in the measures of diversity used by ecologists. In other words, these diversity indices are based on an underlying value statement on the part of community ecologists, that every species in a sample is equivalent to—and interchangeable with—every other species in the sample. In contrast, a central value of conservation decision makers is that all species are not equal—effective conservation requires a knowledge of the identities of species. Conservation decision makers value species that are endangered rather than common; most conservation organizations would choose to devote their resources to protecting a single population of a threatened species rather than to a large number of populations from an assortment of common species. Also valued by conservation decision makers are those species that contain potentially valuable chemical compounds; those that are endemic to a particular region; those that have particular symbolic worth to the people of the region, or even the people of the world, as with the African elephant or the giant panda; and those species, like the tuatara, that have no close relations among other extant species. Moreover, both community ecologists, despite their reliance on species richness as the primary tool for quantifying diversity, and conservation biologists place particular value on "keystone species," those species that play critical roles in their ecosystems. The removal of these species from their ecosystem can have far-reaching effects, such as causing serious changes in population size and viability in other species that rely on them for food or shelter or compete with them for resources. To conservation decision makers, species are not fungible units.

The second issue, the fact that ecologists and conservationists have different underlying reasons for being interested in diversity and biodiversity at all, is well illustrated by the following quotations, the first from Schluter and Ricklefs, in the opening chapter of their edited volume *Species Diversity in Ecological Communities:*

> The most fundamental data of diversity are the numbers of species in different places. Ecologists have discovered relationships between these data and latitude, climate, biological productivity, habitat heterogeneity, habitat complexity, disturbance, and the sizes and dis-

tances of islands. . . . Several of these relationships have suggested mechanisms that might regulate diversity; a general and comprehensive theory of diversity must account for all of them.[10]

As can be seen from this quotation, diversity for community ecologists is simply one factor among many, all of which are employed for the purpose of understanding evolutionary and ecological processes. Diversity, for community ecologists, may be either a dependent variable to be explained in terms of independent variables such as latitude and climate, or an independent variable used to explain other variables, such as stability (a term with its own long history of problematic definitions). On the other hand, for conservationists such as Noss and Cooperrider, biodiversity is a basic good, an essential entity to be preserved, rather than a variable in an intellectual exercise, as this quotation from their book *Saving Nature's Legacy* makes clear:

> The fundamental belief of conservation biology is that biodiversity is *good* and should be conserved. The mission of conservation biology, then, is to conserve as much of global biodiversity as possible and to allow evolution to continue generating biodiversity.[11]

Biodiversity is frequently viewed by the conservation community as a key to the future well-being of humanity and is increasingly viewed as a critical component of sustainable development, as was made clear in the quote from the Convention on Biological Diversity at the start of Chapter 2.

Thus, despite the obvious and confusing similarity of the terms *diversity* and *biodiversity*, these two terms clearly mean very different things to ecologists and conservationists, and care must be taken to distinguish between them. The current state of the literature is such, however, that the two terms, *diversity* for the ecologists and *biodiversity* for the conservationists, are still occasionally confused, as can be seen in the following quotation from a paper by Pressey et al., which advertises a discussion of *biodiversity*, but quickly switches, without distinguishing the terms, to a discussion of the community ecologist's concept of *diversity*:

> **Definitions of *biodiversity* and their use in reserve selection**
>
> There are several ways of defining *biodiversity*, which is itself a rubric to cover all of nature's variety. For any measure of *diversity* to be useful for conservation evaluation, it should be able to represent both alpha, or local, diversity and beta diversity—variety among areas (gamma diversity at larger spatial scales). . . . One of the simplest measures of *diversity* is species richness, a count of the number of species. Ecologists interested in community structure have also included the relative abundances of species in composite measures.[12] [Emphasis added]

The premise behind the use of this concept of diversity by Pressey et al. is that nature reserves should be selected to include a maximal number of species. And *in the absence of all other knowledge,* we would agree with this statement. However, as soon as particular knowledge of other conservation values, such as those we outlined above—endangeredness, presence of valuable chemical compounds, symbolic worth, or taxonomic isolation—becomes available with regard to particular species, the usefulness of species richness becomes substantially weakened. Moreover, as the values of conservation decision makers change from protection of species diversity to protection of diversity among ecosystems, any definition of diversity that depends on species numbers alone will unnecessarily constrain conservation policy. For our last example of the problematic relationship between diversity and values, we turn to one aspect of ecosystem diversity, a scenario from one of our class field trips that has raised provocative questions among our students concerning their own values and the many ways of finding diversity in the natural world.

THE LANDSCAPE MATRIX IN VERMONT

As we travel with our students through the hills of northeastern Vermont in early October, we are struck by the vivid fall colors—the bright reds, oranges, and yellows of the sugar maples and the burgundies of the white ashes against the deep greens of the spruces and firs. At times, we drive down roads enclosed by the comforting overhang of the forest on both sides; at other times, the road opens up to reveal a patchwork of farms and forest. When we are high on the hills and the trees are not hard by the road, we can see around for many miles. Our views include vast forests whose colorful leaves are brightened by the sun, and emerald rectangles of pastures being grazed by Holsteins and Jerseys, separated by windrows of trees. In the midst of these fields and in the yards of the farmhouses are spreading, open-grown elms, sometimes alive and golden yellow, sometimes dead, revealing their vaselike skeletons.

Most of New England was deforested for farmland by the early to mid-nineteenth century. It was a hardscrabble life of dirt farms and sheep grazing. The land was rocky, so clearing took inordinate time and effort, and the topsoil was thin. When the pioneers discovered that the land of the tall-grass prairie to the west had deep, rich topsoil, and that clearing it was much easier, many New England farmers left their homes to plow the fields of Ohio, Indiana, Illinois, Iowa, and points west. In Vermont, the land that stayed farmland went to dairying, and the rest reverted to forest. One hundred years ago, 90% of northern Vermont was cleared land, and 10% was forest. Today, the percentages have been reversed, and with each farm failure, the land moves closer to total forest cover, and reports of the return of mountain lions and wolves become more frequent each year.[13]

For most conservationists, this is a success story. But in fact, as northern Vermont moved from 80% forest to 90% forest and from 20% farmland to 10% farmland, the diversity of the landscape was, in one sense, diminished. And for many people, some of them conservationists, the charm of the landscape was diminished. Total forest cover brings with it a value, wildness, that most conservationists hold dear. Loss of farmland brings loss of other values that many people hold dear. Some are associated with the decline of farming itself, and the loss of diversity of what Hesiod (the author of the epigraph that opened this book) would consider honest ways of making a living. But others are associated with the diversity of ways of appreciating the forest, for the open vistas offer perspectives of the forest that cannot be found deep within it. As Louis Menand notes in the essay "Diversity" with which we opened this chapter, there is a "lesson in diversity, which is that difference—that which is not us—is what makes identity possible."[14]

Can it be that we lose an important way of appreciating the forest when we do not have unforested land to compare it with, to see it from? Do we want a Vermont without dairy farms? These questions have no easy answers, but because of the diversity of backgrounds and viewpoints that we value so highly in our students, our class discussions of these questions during the crisp autumn evenings in Vermont are as vibrant and poignant as the leaves in the next morning's sunlight. The students begin to see the three ways of looking at diversity that we outlined at the beginning of this chapter. They note that there are some kinds of diversity that they ignore: most students do not really care if diversity of dairy cows—Holsteins, Guernseys, Ayrshires, Brown Swiss, Jerseys—on the Vermont landscape goes up or down. They agree that there are kinds of diversity that they all can actively value, such as the diversity of natural ecosystems and native species that they see throughout northern Vermont. But they also acknowledge that there are kinds of diversity that they attend to but actively disvalue. For example, some of the students desire no diversity in land use patterns, preferring to see northern Vermont go back to complete forest cover; this argument places diversity second to ecological integrity. Although other students prefer the patchwork diversity made up by the northern forests and the dairy farms, the interaction of their various positions helps them all see that diversity is a term with no essential content, to paraphrase Menand's epigraph to this chapter, until it is infused by the values of the beholder.

References

1. Louis Menand, "Diversity," in *Critical Terms for Literary Study*, ed. F. Lentricchia and T. McLaughlin (Chicago: University of Chicago Press, 1995), 336.

2. U.S. Department of Agriculture, *Sylvics of North America*, vol. 1 (Washington, D.C.: U.S. Government Printing Office, 1990), 601–603.

3. F.W. Preston, "Diversity and Stability in the Biological World," in *Diversity and Stability in Ecological Systems* (Brookhaven Symposia in Biology, no. 22, 1969), 1.

4. J. R. Karr and D. R. Dudley, "Ecological Perspective on Water Quality Goals," *Environmental Management* 5 (1981): 55–68.

5. U.S. Department of Agriculture, *Apples: Production Estimates and Important Commercial Districts and Varieties* (USDA Bulletin No. 485, 1917).

6. Peter Wynne, *Apples* (New York: Hawthorn Books, 1975).

7. Robert M. May, "Conceptual Aspects of the Quantification of the Extent of Biological Diversity," in *Biodiversity: Measurement and Estimation. Philosophical Transactions of the Royal Society London*, B 345, ed. D. L. Hawksworth (1994), 14.

8. D. Walker, "Diversity and Stability," in *Ecological Concepts*, ed. J. M. Cherrett (Oxford: Blackwell Scientific Publications, 1989), 115.

9. International Council for Bird Preservation, *Putting Biodiversity on the Map: Priority Areas for Global Conservation* (Cambridge: International Council for Bird Preservation, 1992), 3.

10. Dolph Schluter and Robert E. Ricklefs, "Species Diversity: An Introduction to the Problem," in *Species Diversity in Ecological Communities*, ed. Robert E. Ricklefs and Dolph Schluter (Chicago: University of Chicago Press, 1993), 2.

11. Reed F. Noss and Allen Y. Cooperrider, *Saving Nature's Legacy* (Washington, D.C.: Island Press, 1994), 84, 86.

12. R. L. Pressey, C. J. Humphries, C. R. Margules, R. I. Vane-Wright, and P. H. Williams, "Beyond Opportunisms: Key Principles for Systematic Reserve Selection," *Trends in Ecology and Evolution* 8 (1993): 124–128.

13. Russ Spring Jr., personal communication, October 1995.

14. Louis Menand, "Diversity," in *Critical Terms for Literary Study*, ed. F. Lentricchia and T. McLaughlin (Chicago: University of Chicago Press, 1995), 351.

5

Mapping the Patterns of Biodiversity

Not only is it easy to lie with maps, it's essential. To portray meaningful relationships for a complex, three-dimensional world on a flat sheet of paper or a video screen, a map must distort reality. As a scale model, the map must use symbols that almost always are proportionally much bigger or thicker than the features they represent. To avoid hiding critical information in a fog of detail, the map must offer a selective, incomplete view of reality. There's no escape from the cartographic paradox: to present a useful and truthful picture, an accurate map must tell white lies.

Mark Monmonier, *How to Lie with Maps*[1]

Every year on our Vermont field trip, we take our students to Barr Hill. After a short climb to an open area on the side of the hill, we are able to view a large expanse, hundreds of square kilometers, including areas that we have been studying at ground level during the previous few days. Our views of the landscape, from this broader perspective, reveal patterns quite different from those we have been seeing on the ground; like mapmakers, we have a "selective, incomplete view of reality." But by choosing this particular perspective, this specific incomplete view, we can perceive features of the landscape that we were unable to recognize from within the forest; we can now see clearly marked bands of different types of vegetation on distant hillsides—dark green bands alternate with multicolored bands of red, orange, and yellow. Choosing next to look at closer hillsides, selecting yet another view of reality, we can see that the bands consist of different types of trees, that the dark green areas contain conifers and the warm-toned areas consist

Figure 5-1.
Views from Barr Hill in Vermont. Views of distant and nearby hillsides reveal different types of information; the distant scenes reveal broad patterns, while the closer ones show more detail about the trees. (Photographs by Dan L. Perlman) (See Color Plates 8 and 9 for other views from Barr Hill)

of deciduous trees in their autumn displays (see Figure 5-1). Moreover, we see that on these closer hills the patterns are not quite as distinct as they appeared on the more distant hills; a zone of mixing between the conifer and deciduous zones contains elements of both, and an occasional patch of green can be seen, like an island, in the multicolored deciduous zone. Selecting an even closer view, looking at the hillside on which we are standing, we notice differences within each patch, finding paper birches and sugar maples in the multicolored areas and spruces and firs in the green.

Throughout the process, beginning with our view of the most distant hillsides and ending with our view of Barr Hill itself, *we* have selected the view, the section of the landscape on which to focus. Like the mapmaker, we choose to emphasize certain features of the landscape even as we actively choose to ignore others. As a result of our choices, we are able to perceive different patterns across the landscape with each shift in vision, as we might by viewing a variety of different maps of a single region one after another. There is a rich and complex reality "out there" that we cannot perceive at a single glance. But by selecting specific views, by *valuing* one view above all others, and then valuing a different view, we are able to gain a more complete picture of the landscape. In a very real sense, what we perceive is constrained by what we value; if we do not actively choose to view distant hillsides, we will not see the patterns of the broader landscape. But if we ignore the hillside on which we stand, we cannot appreciate the finer variations in the broader patterns.

Throughout this chapter we look at maps: maps of geographic landscapes; maps showing changes in time; and maps depicting the distribution of biodiversity across the Tree of Life. As we did on the side of Barr Hill, we continually make choices about what to view, what to present, and what to ignore—we express our values by deciding which information is important to present. As we attempt to create useful maps, we must necessarily tell lies, but we hope that by acknowledging the lies we tell, we will not mislead our readers, and that by presenting enough maps, the underlying patterns of the landscapes will not be obscured by the necessary incompleteness of any one particular map.

THREE DIFFERENT LANDSCAPES

As we sit on the side of Barr Hill, Russ Spring, a naturalist and eco-tourism lodge owner and our guide to the area, introduces us to the patterns of biodiversity in a landscape quite different from the geographic one that we have been viewing; he tells of changes in biodiversity over a second landscape, across the temporal landscape—the landscape of history. He describes the typical patterns of succession found on hillsides such as those we see before us, telling us of the species and communities that are first to invade a burned or cleared area and those that replace them. From the patterns of different forest types that we see on the hillsides, he deciphers events of decades ago. As he talks, he pushes further back into time, telling of how this lush, forested landscape was almost completely denuded for sheep farms a century ago, and how it was under a thick layer of glacial ice just 12,000 years ago. With his vivid descriptions he creates a series of maps for us, with each map portraying patterns of biodiversity that spread across the geographic landscape at a different time in history.

As we leave our hillside lookout and walk into a nearby forest, we consider patterns of biodiversity in a third "landscape," the taxonomic landscape of the Tree of Life. As we explore this taxonomic landscape, we begin mapping it in our minds, creating a map of where especially important "regions" of biodiversity are located. Within the landscape we pay special attention to the woody plants, small flowering herbs, and insects—and create especially detailed areas of these regions of our map. In addition, we discuss the ecological importance of fungi and bacteria, while acknowledging that we (both as individuals and as a species) know little about these parts of the taxonomic landscape. As we create our map of this landscape, we change scales frequently, as we did in mapping the geographic and temporal landscapes. At times we discuss entire kingdoms, such as the kingdom Fungi, while at other times we consider diversity within lower taxonomic units; we discuss the diversity within orders (such as the beetles), families (such as the pine family), and genera (such as the birch genus *Betula*).

Just as we do with our class on Barr Hill, we attempt, in this chapter, to understand and appreciate biodiversity across three different "landscapes." As in the field, we begin by discussing and mapping biodiversity as it occurs in the geographic landscape—this is the landscape that we can see as we stand on Barr Hill. Next, we view the temporal landscape and consider the ways in which biodiversity has changed over time. This landscape is not as easily mapped as the geographic landscape, but with the study of fossils and a variety of other techniques, we can begin to view historical changes in patterns of biodiversity. Finally, we turn to the taxonomic landscape, discussing those portions of the taxonomic map where much of the diversity of life resides, while acknowledging that large areas are still marked "Terra Incognita." In this chapter we do not use the term *landscape* in the technical sense that landscape ecologists use it, meaning a land area containing a variety of ecosystems that is smaller than a region.[2] Rather, we take the dictionary definition of the word *landscape*, "an expanse of natural scenery seen by the eye in one view,"[3] and extend it beyond its obvious reference to the geographic landscape to include a view of history (the temporal landscape) and a view of the Tree of Life (the taxonomic landscape). Throughout the chapter we will describe the distribution of biodiversity in terms of maps. Some of these maps will look familiar, as they show the distribution of biodiversity across geographic regions, while others will be maps in a metaphorical sense, showing biodiversity across the temporal and taxonomic landscapes.

A BRIEF WORD ABOUT MAPS

In his book *Sylvie and Bruno Concluded*, Lewis Carroll describes another kind of map, one drawn of the German countryside. The difference between this map and all others is that it is drawn on a scale of one inch to the inch. It shows every detail, but it can never be used, because whenever it is unfolded, it keeps sunlight from reaching the crops.[4] The maps that we use and discuss in this chapter are, like all maps, abstractions of reality. Unless a map

Color Plate 1.
Passenger Pigeon. (*Ectopistes migratorius*). In the United States, populations of the passenger pigeon were larger than those of all other bird species combined until the middle of the nineteenth century. Still, this element of the ecological legacy did not survive to our times. (Photograph by Glenn Adelson, from the collection of Museum of Comparative Zoology, Harvard University. © President and Fellows of Harvard College)

Color Plate 2.
The Tropical Rainforest. Has the overwhelming diversity of tropical rainforests such as this site in Costa Rica dulled our analytical approaches for assessing biodiversity? (Photograph by Dan L. Perlman)

Color Plate 3.
Impatiens wallerana. This lovely flower is not native to the forests of Monteverde, Costa Rica, although it grows there in great abundance. In what ways does its presence in those forests add to, or detract from, the region's biodiversity? (Photograph by Dan L. Perlman)

Color Plates 4, 5, 6.
Scale Insects (Family Margarodidae). Most measures of diversity and biodiversity consider the individuals of a given species as interchangeable. To what extent are the strikingly different males and females of this species interchangeable? Color Plate 4: Male. Color Plate 5: Female. Color Plate 6: Male and female mating. (Photographs by Dan L. Perlman)

Plate 4

Plate 5

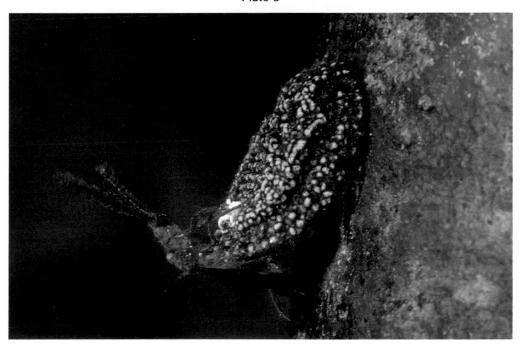

Plate 6

Color Plate 7.
Caterpillar and Butterfly Shadow. Dramatic change during development is another characteristic not accounted for by measures of diversity and biodiversity. Should we distinguish between caterpillars and adult butterflies in our measures of diversity? (Photograph by Dan L. Perlman)

Color Plates 8 and 9.

Barr Hill, Vermont. Different patterns of biodiversity emerge when different scales are considered, as can be seen from a site that we visit each year with our students. Color Plate 8: Distant hillsides. Color Plate 9: Nearby hillside. (Photographs by Dan L. Perlman)

Plate 8

Plate 9

Color Plates 10 and 11.

What you see depends on where you look. These two photographs were taken moments apart. The only difference in procedure between the two photographs was to change focus from a nearby pokeweed (*Phytolacca americana*) to a distant nature photographer. (Photographs by Dan L. Perlman)

Plate 10

Plate 11

Color Plates 12, 13, and 14.
Mary Dunn Pond. Ecological succession could change this coastal plain pond, home to species of special con
like the Plymouth gentian and the thread-leaved sundew, into an oak-pitch pine forest like most of the surroun
area. Can the way we define the term *ecosystem* affect the likelihood of success of our conservation efforts? Co.
Plate 12: Plymouth gentian (*Sabatia kennedyana*; photograph by Glenn Adelson). Color Plate 13: Thread-leaved
sundew (*Drosera filiformis*), a carnivorous plant system shown having captured a damsel fly (*Enallagma* sp.)
(Photograph by Glenn Adelson). Color Plate 14: Mary Dunn Pond. (Photograph by Dan L. Perlman)

Plate 12

Plate 13

Plate 14

Color Plate 15.

The "Bullpen" Orchid (*Stellilabium bullpenense*). This tiny orchid, just a few millimeters across, is known only from one site, the tree-studded Costa Rican pasture in which it was first discovered in 1989. How much can we learn, and how much inspiration can we draw, from elements of biodiversity that remain to be discovered or studied in detail? (Photograph by Dan L. Perlman)

represents the landscape at a 1:1 scale, like Carroll's fabulous map, a map-maker must choose one or a few attributes of the real world on which to focus (e.g., roads, topography, distributions of species, habitat types), while ignoring all others. Through the abstraction of the mapmaking process, one can highlight specific features of the landscape and display patterns of these features. The necessary corollary of the abstraction process is that most attributes of the landscape are left out of any single map. So, when using a map, it is important to understand the mapmaker's goals and to be wary when using the map for purposes other than those for which it was made. A map showing patterns of ecosystem types, for example, is not a good tool for navigating from one place to another, although it is invaluable for understanding the global distribution of biodiversity (see Figure 5-2). Being clear about the reasons that a map was created helps one understand the strengths of the map, as well as its weaknesses, which in turn enables one to know when to look for a different map—a different tool—to help answer a given question.

Along with understanding the mapmaker's purpose in creating a certain map, the map user must be aware of the resolution or scale that was employed in gathering data for the map, as well as the resolution at which the data are presented on the final map. For example, a mapmaker mapping the ecosystems of a 2000-square-kilometer region might choose to gather data with a resolution of 100-meter-wide plots, 1-kilometer-wide plots, or even 10-kilometer-wide plots. A pond 90 meters in diameter shows up on a map with 100-meter resolution, but it would likely not appear on maps created with the coarser resolutions. As a general caveat for map users—do not be seduced by a map's title alone. Not only must the subject matter of a map and its area of coverage be useful, but its level of resolution must also be appropriate to one's needs. To someone hoping to locate tiny, temporary vernal pools, an ecosystem map created at a resolution of 10-kilometer patches will be of no use.

A few years ago our class saw a powerful real-life example of the importance of these basic mapping issues. During our annual visit to Costa Rica, we saw a public presentation that was part of the creation of a master plan for the Monteverde Conservation League. A pair of outside consultants had acquired satellite photographs of the region to help in the planning process, and were using these as the centerpiece of the plan. Unfortunately, although the resolution of the commercially available photographs was reasonably fine (each pixel or dot in the digitized photograph represented an area 30 meters by 30 meters), the consultants were only able to distinguish five types of land use in the region: primary forest, secondary forest, pasture, lake surface, and the denuded side of a volcano. The inability to distinguish among different types of primary forest was especially problematic; in the Monteverde Cloud Forest Preserve, just one portion of the region being mapped, forest ecologist Robert Lawton has recognized six distinct forest types—and several other types exist in the broader region.[5] Because it is exactly such information that is critical in conservation planning, the map as created was of little use.

Mapping is a metaphor that we use to illustrate the point that, despite apparently straightforward definitions of biodiversity, employing the concept in practice is a complex task. When a community ecologist claims to

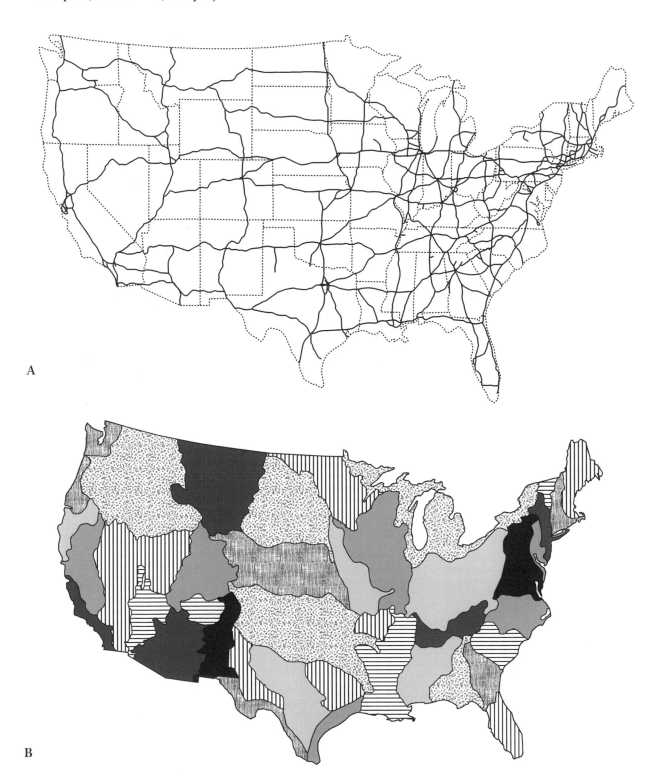

Figure 5-2.
Two maps of one region. A single region may be mapped in different ways, for entirely different purposes, as can be seen in this pair of maps of the continental United States. (A) This roadmap shows major highways, while (B) this map shows "ecosystem units" as defined by the U.S. Fish and Wildlife Service. (U.S. Government Accounting Office, *Ecosystem Management* [GAO/RCED-94-111, 1994] 46)

A

B

measure biodiversity by counting species in an ecosystem, or when a systematist claims to measure biodiversity by quantifying branch points on a cladogram, we must appreciate just how little of the entire concept of biodiversity such a procedure encompasses.* As we demonstrate throughout this chapter, in order to know something about specific assemblages of biodiversity in the world, one must make specific choices about how and where to study in the full realization that those choices provide one with an incomplete view. Our understanding of the Earth's biodiversity is highly dependent upon choices we make in studying that biodiversity, and our range of choices is not as limited as sometimes appears in the literature.

MAPPING BIODIVERSITY ACROSS THE GEOGRAPHIC LANDSCAPE

During the course of a school year, our class does a great deal of traveling by a number of different means, at a variety of paces. Below, we present four means by which we travel, each approximately one order of magnitude slower than the preceding one, and each correlated with a different scale of biodiversity—biomes, ecosystems, dominant species, and smaller species. We then provide a fifth mode of transport, correlated with the genetic scale of biodiversity, which is different from the other four in that proteins or DNA, not humans, are moving. The rate of movement at this scale is many orders of magnitude slower than the slowest of the others. Each mode of transport gives us a different view of the physical world, and each enables us to create a different map of the world's biodiversity. What does each of these maps show us about the distribution of biodiversity across the geographic landscape, and what does each map leave out?

Biome Scale/Flying Pace

Flying 10,000 meters above the surface of the Earth in an airplane, moving at a speed of greater than 600 km/hour, one can see broad patterns on the land and in the oceans (see Figure 5-3). Cities and roads are clearly visible, and agricultural areas can be distinguished from surrounding natural vegetation. Over more natural areas, those least affected by humans, one can distinguish lakes and rivers from surrounding lands, grasslands from savannas and open woodlands, open woodlands from forests, and evergreen forests from deciduous ones. Over the oceans, shallow continental shelves, coral reefs, and deep ocean each have their own characteristic patterns.

At this scale, we can view biological diversity at the level of *biomes* (see Figure 5-4). These great landforms, which are distinguished on the basis of their primary vegetation forms, may stretch for hundreds or thousands of kilometers when uninterrupted by human activities, as do the arctic tundra and a few regions of tropical rain forest. In areas with a great deal of topographic variation, multiple zones of vegetation that correspond to different biomes can be seen close together. For instance, as we fly over tall mountains, we may notice that conifer forest appears at high elevations in a

*See pages 97–99 for a discussion of cladograms.

Figure 5-3.

View of the Earth's surface as seen from an airplane. From this altitude, only the grossest natural features are discernible, such as major bodies of water, mountains, and differences among biomes. (Photograph by Dan L. Perlman)

region that consists largely of deciduous forest. At very high elevations, the forests may disappear altogether, to be replaced by alpine areas that resemble the tundra of the far north. Flying over the countryside, we can see patterns of biodiversity quite clearly, but for the most part, the map we could create from such a view would only show patterns at the level of the biome. From the airplane window, moving rapidly, details such as ecosystems and communities within biomes, not to mention individual species and organisms, simply cannot be seen.

If we look at biome-level maps of the world, we find that in areas with little topographic relief, different parts of the globe do not vary drastically in the number of biomes that they contain. In other words, nonmountainous temperate and tropical regions do not differ in the amounts of biodiversity that they hold—*at the biome level* (see Figure 5-4). In 1992, the World Conservation Union, or IUCN, a collection of governmental and nongovernmental organizations, "established a goal of having 10 percent of each of the world's major biomes (a broad regional ecological community) protected by the year 2000,"[6] so this is a level of biodiversity that is given serious consideration by conservationists.

However, it is not clear exactly what such a proposal would entail. As biogeographers C. Barry Cox and Peter D. Moore have noted, "There is no real agreement among biogeographers about the number of biomes in the world."[7] Who is to assemble the list of biomes that are to receive protection, and how the list is to be constructed, are still open questions.

Figure 5-4.

Biomes. Biogeographers divide the Earth's terrestrial biodiversity into a relatively few biomes, as shown in this map. (C. Barry Cox and Peter D. Moore, *Biogeography*, 5th ed. [Oxford: Blackwell Scientific Publications, 1993])

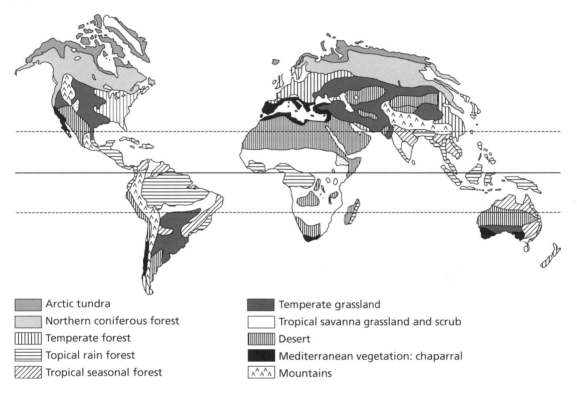

Arctic tundra
Northern coniferous forest
Temperate forest
Topical rain forest
Tropical seasonal forest

Temperate grassland
Tropical savanna grassland and scrub
Desert
Mediterranean vegetation: chaparral
Mountains

Ecosystem Scale/Driving Pace

Only rarely does our class travel by air—when we go to and from Costa Rica—so we do not often get to see the world's biodiversity in terms of biomes. Frequently, however, we travel by bus or van, moving at about 60 km/hour. What kinds of patterns are visible at this pace, and what features of biodiversity can be mapped at this rate of travel?

Imagine that you are driving with our class on a dirt road cut through a fairly natural area. Moving in a van at this pace, we can appreciate finer differences than we could from the airplane. For example, we can distinguish patches of forest that are of different ages—we can see that some parts of the forest have shorter or slimmer trees than other parts, and we can pick up differences in the proportions of conifers and deciduous trees in a temperate forest, or we may be able to recognize a few distinctive pioneer or old-growth species. One can get a view of where a shrub layer exists in the forest, and one can sometimes see whether a distinct herb layer shows up as a carpet of green in the forest, or whether the forest floor is just covered with brown, decaying vegetation. If we were to drive through grassland, we could see the difference between tall-grass and short-grass prairie. If we were lucky, we might see a large mammal or bird, but for the most part our ability to discriminate among elements of biodiversity is limited to the community/ecosystem level. Even here, though, we cannot fully discriminate among all ecosystem types, because appreciating some of these differences requires the precise identification of species to identify certain habitats.

If we were to map the world moving at driving pace, various patterns of ecosystem and community diversity would become evident in different parts of the globe. In regions with little variation in topography, climate, and soils, one can travel for hours at this speed without seeing much variety in ecosystem types. A good example of this is an arctic region where one sees virtually no variation in ecosystem type other than dry land tundra and the wet bogs and fens. In such areas, one finds that the biome has little variation at the finer ecosystem level.

In contrast, areas with one or more underlying important environmental gradients can display a wide variety of ecosystems and communities. Variations in elevation, climate (such as temperature, total annual precipitation, and the pattern of precipitation during the year), soil type, hydrology, presence of surface water, and patterns of disturbance (such as fire and storm frequency and intensity) all contribute to variation at the ecosystem scale. Such factors frequently are not independent of each other, as when variation in elevation and slope on a mountainside affects temperature, growing season, rainfall, and disturbance (see Figure 5-5). When our class makes its annual visit to Monteverde, Costa Rica, we are able to study several distinct plant communities in an area of just a few square kilometers. This abundance of communities results from the interactions among a steeply varying topography, significant variations in precipitation, and disturbance regimes that include landslides and powerful winds and storms. The forest straddles the Continental Divide, which is raked by the warm, moist trade winds from the Caribbean several months each year; the portions of forest on the windward slope are very wet, while the leeward forest just 2 to 3 kilometers below the Divide receives far less rain—and contains a different plant community. At the very top of the ridge, along the Divide itself, much of the forest's water input is in the form of condensing mists, as the forest is bathed in clouds much of the year. In addition to these patterns of topography and rainfall, variation in soils leads to further differentiation in community types. If we could drive through the forests of Monteverde at 60 km/hour (not likely, given the steep terrain and frequency of downed limbs and treefalls!), in some areas we could see distinct communities flash past at a rate of more than one per minute, whereas we could drive for hours across tundra or temperate steppe grasslands without seeing more than two or three distinct communities.

Dominant Species Scale/Walking Pace

Decamping from our bus or van, we can walk through one of the ecosystems that we have driven through and flown over. Moving at 6 km/hour, we have time for occasional stops. We can get a feel for the species composition of the ecosystem through which we are walking, picking out some of the larger and most common species. At this pace, we can create a map showing the number of species (at least the large, easily observed ones, such as trees) known to exist in each ecosystem. Using such a map, one can compare the number of species found within one ecosystem with the number found within another and gain an appreciation of how species richness varies among sites.

Figure 5-5.

Increased diversity of communities. One can find several different ecological communities clustered together in mountainous areas where elevation, slope, exposure, temperature, and disturbance all vary considerably, as shown in this diagram of plant community distributions on an idealized west-facing slope in the Great Smoky Mountains. Communities: BG, beech gap; CF, cove forest; F, Fraser fir forest; GB, grassy bald; H, hemlock forest; HB; heath bald; OCF, chestnut oak-chestnut forest; OCH, chestnut oak-chestnut heath; OH, oak-hickory; P, pine forest and heath; ROC, red oak-chestnut forest; S, spruce forest; SF, spruce-fir forest; WOC, white oak-chestnut forest. (After R. H. Whittaker, "Vegetation of the Great Smoky Mountains," *Ecological Monographs* 23[1956]: 41–78)

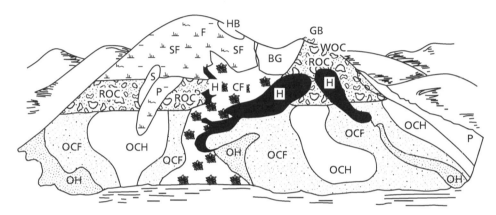

Looking at a global map depicting the number of tree species within ecosystems, we find that some temperate forests, such as those of the northern boreal regions, contain only a few tree species. Even the richest temperate forests, such as those of the southern Appalachian Mountains in the United States, rarely contain as many as 25 canopy tree species. In contrast, tropical forests often contain 100 canopy tree species in a single hectare, an area about the size of two soccer fields, and may hold as many as 300 species per hectare.[8] Viewed at this pace, one that allows for the identification of most tree species, one consistently finds many more species in the tropics than in the north temperate zones, which in turn contain more species than the arctic zones. Southern temperate regions, which are less studied than the north temperate, often contain large numbers of tree species, although generally not as many as tropical regions.

Further distinctions in the number of tree species per region can be made as well. Isolated oceanic islands typically have lower numbers of species than continental regions of similar climate and area, although many of the species found on such islands are endemic (i.e., not found elsewhere). Within the north temperate zone, Asian forests typically have many more tree species than comparable North American forests, which in turn contain more tree species than European forests. It has long been argued that the east-west orientation of the major European mountain ranges prevented the southward migration of many species during glaciation events, leading to the extinction of many tree species in the region, while the north-south orientation of the North American mountain ranges was not an impediment to southward migration, enabling that continent's flora to remain more intact.

In contrast, most of East Asia's forests have remained free of glaciers, while being repeatedly enriched by immigration of species from neighboring subtropical forests.[9]

Smaller Species Scale/Crawling Pace

Once we are in the field with our students, we move slowly. We may walk 600 meters in an hour, although we often do not reach such a rapid pace. In fact, we frequently ask our students to spend an hour or two examining and mapping an area of 100 square meters, and sometimes we ask them to study an area less than 1 square meter. In these situations it might be said that we move at a crawl, at a rate of 0.06 to 6 meters per hour.

In this range of paces and scales, individual trees become major features of a map. In fact, at the finest scale we study, individual leaves of the elephant ear (*Xanthosoma*) and *dos amigos* (*Gunnera*) plants we see in Costa Rica, which can be up to a meter across, could cover several study plots entirely.

Our study plots are so small that they cannot include a full sampling of the species found in a given ecosystem; at these scales, our maps contain only a fraction of the species present. However, moving at this pace we begin to appreciate the diversity of species other than trees; we can sample widely from the soil fauna and flora, from the forest floor herbs and shrubs.

Because the characteristics used to distinguish these small species are often quite subtle, we teach our students the identifying characteristics of taxonomic units higher than species, frequently describing genera, families, and orders. Regardless of the taxonomic level or group under consideration, as a general rule, most tropical regions have more species than most temperate regions. For example, as mentioned in Chapter 2, the Monteverde region hosts over 600 species of butterflies, some of which are year-round residents, while others are migrants from the dry lowland forests on the Pacific side of the country, and yet others just pass through as they migrate from the dry forests to the wetter forests of the Caribbean side. This number of species is all the more impressive when compared with the roughly 440 species found in eastern North America and 380 species of Europe and the North African Mediterranean coastal region.[10] A few groups, including salamanders, aphids, and conifers, however, run counter to the general trend; in these groups, temperate regions have more species than tropical regions. Table 5-1 lists several comparisons of species richness between temperate and tropical regions. Species richness by itself omits most of the information that conservation decision makers consider vital: whether the species is endangered, endemic, or taxonomically isolated; whether it contains important medicinal compounds; and whether it holds symbolic value for humans. We present information about species richness in this chapter, not to suggest that it should be the grounds for conservation decision making, but as illustrative of one of the ways of mapping biodiversity.

As with tree species, at these scales, the species richness of south temperate areas stands out, especially those regions with a Mediterranean climate (cool, wet winters with freezing temperatures only rarely, and hot, dry summers), such as southern and southwestern Australia, the Cape region of South Africa, and Chile. These areas contain spectacularly large numbers of plant species, many of which are endemic to the regions. California and the

Table 5-1. Comparisons of species richness between temperate and tropical regions, a few selected examples.

Group	Tropical Regions	Temperate Regions
Ants[a]	275 species in 8 hectares in Peru	100 species in all of New England; approximately 650 species in all of the United States
Snakes	56 species in 1500 hectares in Costa Rica (La Selva Biological Station)[b]	91 species in the eastern United States and Canada[c]
Birds	more than 400 species in 1500 hectares in Costa Rica (La Selva Biological Station)[d]	650 species breeding in all of the United States and Canada[e]

[a]Stefan Cover, personal communication, October 1995.
[b]H. W. Greene, personal communication, July 1990.
[c]Roger Conant and Joseph T. Collins, *A Field Guide to Reptiles and Amphibians*, 3d ed. (Boston: Houghton Mifflin, 1991).
[d]T. C. Whitmore, *An Introduction to Tropical Rain Forests* (Oxford: Oxford University Press, 1990), 59.
[e]Paul R. Ehrlich, David S. Dobkin, and Darryl Wheye, *The Birder's Handbook* (New York: Simon & Schuster, 1988).

Mediterranean basin, the two similar zones in the Northern Hemisphere, also have large numbers of plant species, including many endemics.[13]

At these scales, our students observe variation and diversity that are a result of microhabitat gradients—essentially, they see beta-diversity (see Box 5-1) expressed on a scale of meters and centimeters. For example, a 1-meter-wide depression at the base of a tiny slope may be slightly wetter than the slope itself and may thus be good habitat for certain species of plants and leaf litter animals that cannot survive on the drier slope itself.

In addition, as with the dominant species of the previous scale, islands tend to have fewer small species than continental areas, but again, islands frequently support large numbers of endemic species due to large adaptive radiations. For example, the islands of Hawaii are home to only 34 species of dragonfly and damselfly, but 27 of these species belong to the endemic damselfly genus *Megalagrion*, a genus found nowhere else in the world.[14] Similarly, the remote South Atlantic island of St. Helena, before it was heavily settled, was covered by forests of woody composites.[15] In other words, on this island, the relatives of the daisies had evolved into an array of endemic tree and shrub species, given the opportunity by a lack of competition from "traditional" woody plant families.

Yet another phenomenon becomes apparent when we move at a crawling pace. Subtle variations among the individuals of a single species become visible, variations that were not visible at more rapid paces. Examinations of the leaf shapes and growth forms of sassafras plants and poison ivy are now possible (see Figure 5-7). In the next section we discuss the most fundamental type of intraspecific variation—genetic variation—which must be studied at an even slower pace.

BOX 5-1.	*Ecologists' characterizations of diversity across the geographic landscape.*

Community ecologists have developed several different ways of characterizing the species diversity of geographic areas.[11] The simplest characterization in this group describes within-community species diversity of a homogeneous site; it is known as alpha-diversity. Alpha-diversity is typically measured in terms of either species richness or species richness combined with equitability, that is, distribution of numbers of individuals among the species counted. The basic question that ecologists want to answer by characterizing alpha-diversity is, Which of these various communities has higher levels of alpha-diversity, and why? To compare sites using alpha-diversity, you calculate the selected diversity measure for each site, then compare the measures to see which comes out higher. Alpha-diversity measures are devoid of context and particulars, and they are not especially useful in conservation planning, except as a rough first pass.

In order to characterize patterns of species diversity across heterogeneous regions, community ecologists employ a concept known as beta-diversity. While there are several ways to calculate beta-diversity, they all answer essentially the same question: To what extent do species turn over (replace one another) between different sites or along an environmental gradient? To assess beta-diversity, one lists the species present at each of the sites being studied, then compares the different lists to see how much overlap exists between sites.

By using beta-diversity, one can observe the impact that variation in underlying environmental conditions (e.g., soil composition and slope) has on the species composition of communities. In regions with low beta-diversity, different sites contain approximately the same suites of species despite any environmental differences that exist there. In contrast, a high beta-diversity value indicates

that environmental variation leads to different suites of species being found in different sites. Note that beta-diversity cannot be calculated for a single site, since it measures the differences *between* sites; as such, it is conceptually quite distinct from alpha-diversity.

Beta-diversity incorporates the context of communities in a fundamental way, since the essence of the measure is the comparison of suites of species at more than one site. The concept is quite useful to conservation decision makers.

When community ecologists assess the species diversity of large, heterogeneous regions that include many distinct sites or ecosystems, they employ measures known as gamma-diversity.

This characterization—which is a counting or index of all the species found in a region—is directly analogous to alpha-diversity. Both are "inventory" diversities; just the scale of the characterizations changes.[12] In contrast, both beta-diversity and delta-diversity (a related but rarely used measure of differences between landscapes) can be described as "differentiation" diversities; they both distinguish among multiple sites or landscapes.

Both alpha-diversity and beta-diversity contribute to the gamma-diversity of a landscape. A region may contain many sites with high alpha-diversity, but if there is little change in the species composition between communities (i.e., if there is low beta-diversity), then the gamma-diversity of the region will not be especially high. In comparison, if the sites of a region each have low alpha-diversity but there is high turnover of species between communities (that is, if beta-diversity is high), then the gamma-diversity of the region may be quite high. Regions with both high alpha- and beta-diversity will have the highest gamma-diversity of all (see Figure 5-6).

Figure 5-6.

Alpha-, beta-, and gamma-diversity. Community ecologists use these three concepts to describe diversity at different scales and of two different types. Alpha-diversity is measured locally, at a single site, as at sites 1 and 2 in the figure. Site 1 has higher alpha-diversity than site 2.

   Beta-diversity measures the amount of change between two sites or along a gradient, as in regions X and Y in the figure. Region Y has higher beta-diversity than region X, as there is a higher turnover of species among the sites in region Y.

   Gamma-diversity is similar to alpha-diversity, only measured over a large scale, such as regions X and Y in the figure. Both alpha- and beta-diversity contribute to gamma-diversity. Region X has high alpha-diversity at its sites, but they are all fairly similar; the region therefore has low beta-diversity and only moderate gamma-diversity. Region Y has low alpha-diversity at its sites, but the sites differ from each other; the region therefore has high beta-diversity, and higher gamma-diversity than region X.

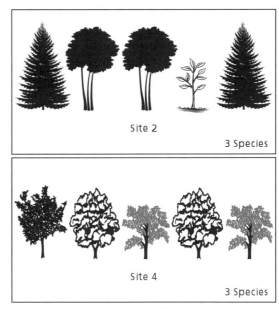

Site 1 — 5 Species

Site 2 — 3 Species

Site 3 — 5 Species

Site 4 — 3 Species

Region X   **Region Y**

Genetic Differences Scale/Electrophoresis Pace

Many of our students work in genetics laboratories, in settings where they are called upon to study genetic differences in populations. In such settings, the students return to the same lab bench day after day, week after week; in a sense, they do not move. What does move, however, are the samples of DNA and proteins that are placed in weak electrical fields. Under the influence of the fields, the samples spread out, revealing different genetic strains.

   Using techniques such as this, an entire new level of biodiversity comes into view. Genetic variation among the individuals of a population or species is generally regarded as a key component of biodiversity, although this facet of biodiversity is not given as much attention as species and ecosystems.

   Relatively little is known about the genetic diversity of most natural populations. Most of the exceptions to this situation are found among crop plants (where wild varieties and relatives are often heavily collected to be stored in seed banks and live voucher collections), among pest and disease

Figure 5-7.

Seeing intraspecific variation. At slower rates of movement, inter- and intra-individual differences become visible, as can be seen in this drawing of sassafras and poison ivy—species that our students get to see during our field trips.

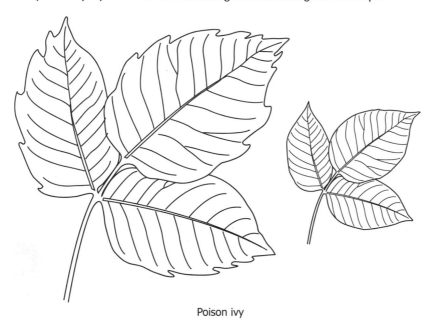

Poison ivy

Sassafras

organisms, and among endangered species. One of the few wild groups that has been studied for genetic variation is the fish family Cichlidae of the Lake Victoria region of Africa. Axel Meyer and his colleagues found that surprisingly little genetic variation exists among the roughly 300 species of cichlids that are found in the lake.[16]

Overall, then, given our lack of knowledge about genetic variation in wild populations, we cannot say much about worldwide patterns of either within-species or between-species genetic diversity, nor can we create a map of this facet of biodiversity. Although tropical regions typically contain many more species than north temperate areas, we do not really know what the relative amounts of genetic diversity of these areas are. Does the pattern of little genetic differentiation among morphologically dissimilar species that Meyer's group discovered among cichlid fish exist in other tropical groups? If so, tropical regions may contain less genetic diversity than one might surmise based on the number of species they hold. One useful map that we can create is a map of the world showing regions where important food crops evolved and where wild relatives of these food crops still exist. Even the most technological of plant breeders can make use of wild organisms as a source of new genetic material to help breed in pest resistance and higher yields, so such regions are of special importance to our species. To a first order of approximation, such a map shows that important regions occur at sites of early, important human civilization, such as the Middle East, China, and the Andes (see Figure 5-8).

No matter where you are on the planet, as you slow down, as you observe more closely, you find more elements of biodiversity. As you slow down from airplane speed to driving speed, you begin to see distinct ecosystems within biomes. A walking pace brings individual dominant species into view within ecosystems, and a crawling pace enables one to see even more species. The even slower technique of electrophoresis allows one to observe genetic variation within species and populations.

A key point here is not whether one sees additional diversity as the pace of travel slows—since of course one would expect to see more detail at a slower pace—but rather that *entirely new categories* of diversity become visible as the pace decreases. Species, for example, are an entirely different category from ecosystems, just as genes constitute a different category from species.

MAPS AND PATTERNS OF BIODIVERSITY OVER TIME

Think back to Barr Hill for a moment, and recall how Russ Spring introduced our students to the history of the site and the larger region. Observing how patches of different tree species were distributed across the hillsides, Russ spoke of the recent history of the area, telling us which spots had burned or been pastureland several decades ago. His understanding of events that took place before his birth was based on his knowledge of oral histories and succession (the patterns by which ecological communities replace one another) combined with his observations of the hillsides before us.

At a slightly longer reach into history, he told us how a century ago the entire region had been covered with sheep and dirt farms—and unhappy farmers. The stories of the frustrations of attempting to grow wheat and

Figure 5-8.

Centers of food crop diversity. Most of the world's important food crops originated in a few geographic areas, sites of early civilizations. Russian botanist N. I. Vavilov noted the patterns of distribution of these crops in what have come to be known as "Vavilov centers." (After W. V. Reid and K. R. Miller, *Keeping Options Alive* [Washington, D.C.: World Resources Institute, 1989], 24)

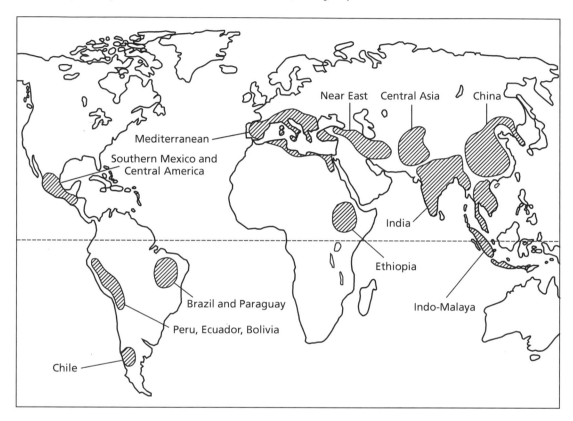

potatoes during a very short growing season in thin, rocky soil had survived in written and oral histories. Russ also told us, based on his knowledge of local geology, that roughly 12,000 years ago this land was virtually devoid of biodiversity, as it was covered by glaciers.

Several general themes emerge from Russ's brief history of biodiversity and land use in northern Vermont. First, note that he used different sources for each type of historical information. This is typical in the reconstruction of biodiversity history—with each change of time scale, entirely new techniques are required. Second, change is frequent and often drastic, even in the absence of human actions (none of the changes to the Vermont landscape before about 10,000 B.C. were caused by humans, since there were none in the area before then, and many of the changes since that time were also nonanthropogenic). Finally, while it may not be apparent from Russ's tale, older periods are much harder to reconstruct and are subject to much more uncertainty in terms of dates and content. In this section we discuss changes that have occurred in biodiversity over time in light of each of these themes. As in the section on geographic landscapes, we employ a variety of different scales.

Note that one can use maps to view the past in two very different ways. First, one can choose a surrogate for all of biodiversity and track what happens to this surrogate over time. A typical choice is to choose a target group, such as marine invertebrates or plants, and to track the number of species or families extant. Alternatively, one can build a temporal series of maps similar to those created for viewing biodiversity in the geographic landscape; flipping through these maps in order will give a moving picture of changes through time.

Months, Years, Decades, and Centuries

These time scales, especially the shorter ones, are accessible to both researchers and ordinary people. Ecologists may study a site for periods that stretch to decades (although this is rarer than one might wish), just as an ordinary person might live in one area, or visit a region many times, over a period of decades. At these time scales, the dominant changes are those wrought by succession and disturbance. As the biological community of a site ages and undergoes succession, it often changes the physical characteristics present—thus creating the conditions necessary for another community to replace it (see Figure 5-9).

Much of the area around Barr Hill was cleared in the nineteenth century for small farms. A few of those farms prospered to the extent that they are still in existence today, but the majority were abandoned. An abandoned sheep or dairy farm in Vermont slowly grows back to become some semblance of the forest it once was, which, in the case of the land around Barr Hill, was the northern hardwoods forest, a forest of beech, sugar maple, yellow birch, ash, and several other tree species. That the abandoned field will return to northern hardwoods we are relatively certain; however, how it grows back will vary, depending on the grazing history of the land and the residual level of human interference. The keen observer who returns year after year to an abandoned northern Vermont field will note a disproportionately large number of sugar maple saplings relative to the number that were in the native forest before clearing and a disproportionately smaller number of beeches. A 100- to 200-year-old successional forest may begin to represent better the species composition of the original forest, but it may still be skewed toward species that, for whatever reason, are more indicative of early succession. Whether the successional forest, when mature, will accurately represent the species composition of the original forest is something we will not know for another 200 years.

But succession does not happen only over the course of years and decades. The microflora and -fauna communities covering a fallen leaf on the forest floor change over weeks and months, as waves of detritus-eating invertebrates, bacteria, and fungi invade the rich resource. With such a shift of perspective and scale, an entire course of succession can be traced on a microscopic landscape.

Disturbance is, in a sense, the opposite of succession—disturbance "resets" the succession clock. A fire, flood, windstorm, landslide, treefall, or human-caused event such as forest clearing can erase (or at least set back) a mature biotic community, making room for earlier successional species and communities. A northern hardwoods forest that eventually replaced an abandoned field in Vermont can be wiped out in a few hours by a fire or

Figure 5-9.

Succession. Few ecosystems remain unchanged over time; instead, they undergo succession. In this figure an abandoned field turns into a secondary forest and eventually a mature forest over the course of decades.

hurricane, and the new open space will again be invaded by the plant species that originally invaded the abandoned field.

Observing at a local scale a patch size that can be altered by a single disturbance event, biodiversity appears to change drastically over time. Species and communities appear and disappear throughout the course of succession, only to appear again with another round of disturbance and succession. If one takes a somewhat larger view, however, and considers a region containing many patches in varying states of succession, one finds that these elements of biodiversity—species and communities—do not really appear and disappear; instead, they move from place to place.

On a global scale, different regions display quite different patterns of succession. Climate and soils have a tremendous influence on the types of communities that can survive in a region; for example, forests generally require significant rainfall during at least part of the year, and rain forests need heavy rains year-round. Moreover, relatively simple patterns of succession, such as those displayed in Vermont, are uncommon in the tropics. Although one may predict that a patch of bare tropical soil will one day be covered with rain forest, the actual species composition of the eventual forest will be far more difficult to predict. Similarly, different regions are subject to different types and frequencies of disturbance.

The processes of succession and disturbance play crucial roles in local patterns of biodiversity, causing species and communities to appear and disappear over the relatively short periods of months to centuries. Over longer time periods and larger geographic scales, however, these processes do little to affect global patterns of biodiversity. Instead, as we shift our focus to longer sweeps of history, we find markedly different processes generating changes in the patterns of biodiversity.

Millennia

As we begin to consider changes in biodiversity over millennia, such as the time since glaciers covered Barr Hill, we discover patterns different from those of shorter time scales. If we consider this time scale from the vantage point of a single site, looking at a series of maps depicting the species and communities present at different times, we see that these elements of biodiversity appear and disappear—often never to return. In North America, for example, following the glacial maximum of roughly 18,000 years ago, tree species and plant communities have migrated northward, in what has been largely a one-way trip (see Figure 5-10).[17]

Although today one can find recognizable communities scattered repeatedly over North America, reconstruction of the past 20,000 years shows that such communities are actually quite labile—migrating, dissolving into their constituent species, and forming entirely new communities at various times. For example, some of the most easily recognized present-day communities of the continent, such as the oak-hickory forest, were "unglued" for much of the past 20,000 years. The major tree species of these forests (many of which were actually oak-hickory-chestnut forests until the chestnut blight early in this century) recolonized eastern North America at drastically different rates, as recorded by the appearance of different types of wind-borne pollen in lake and bog sediments. Whereas oaks reached Rogers Lake in Connecticut (one of the best-studied sites in North America) over 8000 years before the present,

Figure 5-10.

Postglacial tree migration. Individual species and entire ecological communities change their ranges considerably over the course of millennia. As the glaciers receded from North America over the past 18,000 years, new areas were colonized. This figure shows the range expansion of the American chestnut (*Castanea dentata*) throughout eastern North America since the time of the last glaciation. Note that the chestnut reached its northernmost extension a mere 2000 years ago. The lines in the diagram show the date when chestnuts first appeared in an area. (After M. B. Davis, "Quaternary History and the Stability of Forest Communities," *Forest Succession: Concepts and Application*, ed. D. C. West, H. H. Shugart, and D. B. Botkin [New York: Springer-Verlag, 1981], 132–153)

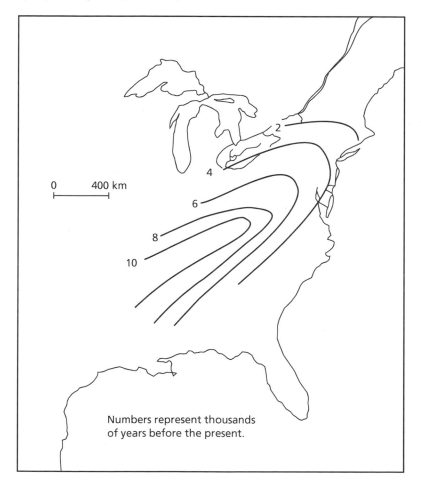

0   400 km

2

4

6

8

10

Numbers represent thousands
of years before the present.

hickories did not reach this site until about 4000 years ago, and chestnuts only arrived 2000 years ago.[18] Further examination of the fossil pollen record indicates that individual tree species had different refugia in which they survived during the periods of glacial maxima.

The great tropical rain forests appear to have covered a much reduced area during the glacial maxima, when dry forests and savannas expanded to cover much of the ground given up by the rain forests (see Figure 5-11).[19] The small refugia into which the rain forests may have been compressed were in many cases widely separated, and many researchers think that such separation led to speciation—in fact, much of the theory of these Pleistocene refugia is based on patterns of species endemism. This theory, however, is still being contested.

Climate is not the only factor that has been altering patterns of biodiversity over the last several millennia; in many ways, *Homo sapiens* has proven to be an equally potent force. From about 12,000 to 9000 years before the present, between 35 and 40 genera of North American large mammals went extinct. Among the species lost were two species of mammoth, one species of mastodon, a giant beaver (the size of a black bear), ground sloths, dire wolves, saber-toothed cats, and the giant short-faced bear (as tall as a moose). *Homo sapiens* arrived on the continent no later than 11,500 years before the present (although we may have gotten here sooner). Although some debate still remains as to whether climate changes or overkill by human hunters caused the large mammal extinctions, the balance of opinion seems to be tipping toward the overkill scenario.[20]

Over the course of the last few millennia, patterns of biodiversity have changed considerably, both locally and globally, with most of the changes being unidirectional (such as the northward migration of tree species in North America and the expansion of rain forest in tropical regions). Only in the last two or three decades have we come to appreciate how drastically species and communities have been affected by climate change and direct human influences. Furthermore, we are now coming to realize that biotic communities are variable in time—groupings that are obvious and seemingly immutable to us today did not exist a few thousand years ago, and we must realize that many of the patterns that appear stable to us are in great flux when viewed over a span of millennia.

Millions of Years, and Beyond

As the interval between past periods and today increases, our ability to find information about the biodiversity that previously existed decreases. Although paleontologists have pieced together a reasonably good picture of past biodiversity by studying the fossil record, gaps in the record in time and space and in various parts of the taxonomic landscape often overwhelm the information that we do have.

As our focus turns to these longer reaches of time, our maps change as well. Instead of employing a sequence of geographic maps to form a moving picture of biodiversity over time, most researchers choose a group of organisms to act as a surrogate for the Earth's biodiversity, and study changes in the group over time. The group most often used as a surrogate is the marine invertebrates—one of the bright spots in the fossil record in terms of quantity and trustworthiness of information available. Because the vast majority of fossils are formed when organisms with hard body parts become embedded in the sediments at the bottom of oceans or lakes, marine invertebrates are perfectly designed and situated to become fossils when they settle to their watery graves. If not disturbed, the remains get covered with sediment and eventually become fossilized. Paleontologists have learned a great deal from the records of marine invertebrates, but

Figure 5-11.
Pleistocene refugia. Ecological communities contract and expand in multiple directions over large areas. (A) At the height of the last glaciation, tropical forests in South America became concentrated into small refugia, (B) as compared with patterns of today. (After G. T. Prance, "Discussion," in *Vicariance Biogeography: A Critique*, ed. D. Nelson and D. E. Rosen [New York: Columbia University Press, 1981], 395–405)

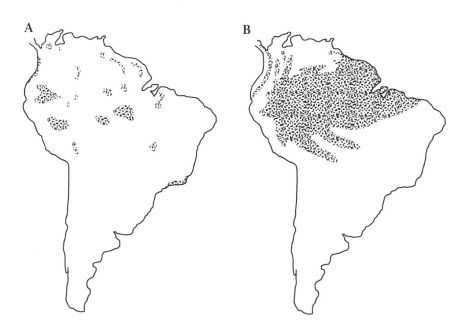

one must also be aware of the ways in which one should not generalize from this distinct, and possibly idiosyncratic, portion of the fossil record. As the authors of *Global Biodiversity* point out, "the fossil record . . . represents a small and highly biased sample of the taxa that have existed—it may represent only one in every 20,000 species that has existed."[21] And while the record of marine invertebrates is reasonably good, plants and vertebrates have left much sparser records.

The marine invertebrates tell a story of the last 600 million years that is quite clear in certain ways. Perhaps the single most important lesson is that diversity among these organisms (with diversity measured as the number of taxa extant) has neither been constant nor has it followed a simple trajectory. There have been periods of relatively fast increase in the number of distinct forms (often measured in terms of the number of families extant) as well as five very sharp drops in the number of forms alive (see Figure 5-12). The drops, which are typically called "mass extinction events," appear to have had several different causes, and took varying amounts of time. The Permian extinction, by far the largest, appears to have been caused by changes in the configuration of the Earth's landmasses, especially the conglomeration of all of the continents into a single mass called Pangaea. With the concomitant loss of a great deal of seashore, climate change, and new volcanic activity, 54% of the families of marine invertebrates went extinct over a 5- to 8-million-year period (as did an even higher proportion of genera and species).[22] By geologic standards, this mass extinction was rapid, but it took more than a thousand times as long to occur as the much smaller but virtually instantaneous 3000-year extinction of the large land mammals of North America. The four other mass extinction events each saw the loss of 15% to 22% of the marine invertebrate families extant at the time, although none compare with the Permian.

Vertebrates show a pattern of mass extinctions that is more or less similar to the marine invertebrate pattern, although fish appear to have suffered three more mass extinctions than marine invertebrates, and tetrapods (i.e., all vertebrates other than fish) have experienced an additional one.[23] The mass extinction in which the dinosaurs were extinguished, at the end of the Cretaceous, hit the tetrapods far harder than fish or invertebrates; of the 89 taxonomic families of tetrapods then extant, 36 were lost.[24]

The pattern for plants is far less clear than that for animals. For one thing, plants do not fossilize as readily as animals, so the record contains more gaps and is therefore more difficult to interpret. It appears that plants generally did not suffer mass extinction events as drastic as the various animal groups, although at the end of the Cretaceous perhaps 75% of the extant plant species perished.[25] Moreover, although paleontologists have been able to reconstruct in part the form of a few ancient plant communities, such as those consisting of ferns and cycads, we know little about the distribution and abundance of such communities.

Figure 5-12.
Mass extinctions. Over the course of the history of life on Earth, several mass extinctions have occurred. This graph depicts the number of families of marine organisms extant over the past 600 million years. The arrows indicate the five mass extinctions of these organisms. (After E. O. Wilson, *The Diversity of Life*. [Cambridge, Massachusetts: Harvard University Press, 1992], 191)

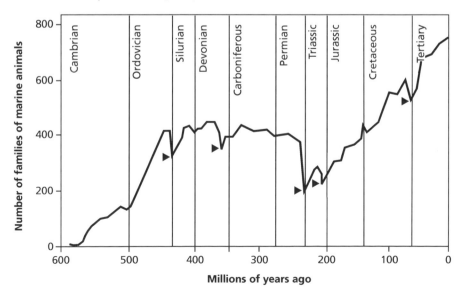

MAPS OF BIODIVERSITY ACROSS THE TAXONOMIC LANDSCAPE

Just as we have used maps to understand where key areas of biodiversity are located in geographic landscapes around the globe, we can use maps to display the location of biodiversity hot spots in the taxonomic landscape. Bear in mind that certain facets of biodiversity cannot be displayed on maps of the taxonomic landscape. It does not make sense, for example, to discuss how ecosystem diversity is distributed through the taxonomic landscape. Instead, in this section we focus largely on the number of taxa located in different parts of the taxonomic landscape, as well as briefly discussing the distribution of genetic diversity.[26] In fact, when reading and thinking about any survey of biodiversity, it is worthwhile to ask which aspects of biodiversity have been included and which have not.

A First Taxonomic Map

Most taxonomic maps of biodiversity employ a scale of species. In other words, they ask, Where are the greatest numbers of species located in the taxonomic landscape? As we can see in Figure 5-13, the largest numbers of species are found in a very few parts of the taxonomic landscape—the insects (especially beetles, but also flies, bees, ants, wasps, moths, and butterflies), other arthropods, and flowering plants account for about 80% of the known species.[27]

Figure 5-13.
A "map" of biodiversity across the taxonomic landscape. Here we map biodiversity according to the number of species *described* to date. (Data from World Conservation Monitoring Centre, *Global Biodiversity: Status of the Earth's Living Resources* [London: Chapman & Hall 1992], Chapter 4)

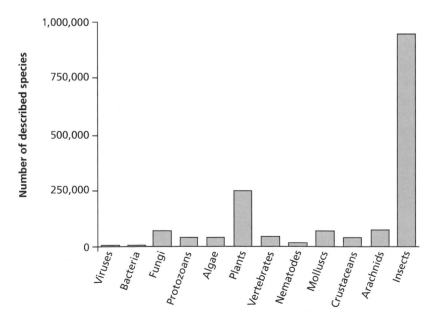

Consider that point for a moment; of the 92 phyla listed by Margulis and Schwartz (see Box 5-2), only *two*, the arthropods (Arthropoda) and the flowering plants (Angiospermophyta), contain about four-fifths of the world's described species. In contrast, all of the vertebrate species combined only account for 2.7% of the total,[30] which is about the number of species found in two *families* of flowering plants combined, the orchids (Orchidaceae) and the composites (Asteraceae).[31]

Notice that thus far we have been discussing the numbers of "described species" rather than "species." A "described species" is one that has received the attention of taxonomists—one or more specimens of the organism have been collected, classified taxonomically, given a new species name, placed in a museum or herbarium, and described officially by publication in a scholarly journal. Such species, however, may be known only from a single specimen, and absolutely nothing may be known of the living organism's ecology or behavior, but at least the species is now "described"—it has a name. Describing a species is an important first step, enabling other scientists to identify and study a species, but it is only a first step. In fact, the vast majority of described species are hardly known beyond the level of a physical description.

A Second Taxonomic Map

In addition to writing about the number of already described species on Earth, biologists often speculate about the number of species that they suspect the Earth actually hosts. Several groups that are barely visible in Figure 5-13 are widely acknowledged by biologists to be largely unexplored, taxonomically speaking. For example, although the number of described arachnid species (class Arachnida includes spiders, mites, scorpions, and their relatives) is only 75,000, and the number of described roundworm species (phylum Nematoda) is only 15,000 to 80,000, many researchers today believe that the *actual* number of species in each of these groups is close to a million (with most of the newly found arachnids being mites), allowing both groups to easily surpass the number of described insect species.[32] But wait—the actual number of insect species is thought to be much higher than the number of described species, with many of the as-yet-undescribed species being found in the canopies of tropical rain forests. For example, entomologists Terry Erwin of the Smithsonian Institution and Nigel Stork of the British Museum have estimated that there might actually be as many as 10 to 80 million arthropod species on the planet.[33] Before proceeding to draw a second map, however, we should consider the various methods by which experts create estimates of the number of unknown species.

As Erwin notes, three methods have been used to estimate the number of undescribed extant species.[34] First, one can search the literature to determine the number of already described species, then add to this the number of species being described each year. Second, one can canvass experts in the field for their best guesses of the probable number of species. Finally, one can extrapolate from data collected in the field to achieve an estimate. Erwin himself chose this final method; he fogged 19 individuals of the tree species *Luehea seemannii* with insecticide and collected samples of the killed insects under the trees. He then counted the number of beetle species in the sample, excluding the weevil species, and made assumptions about the number of weevil species to be expected in the sample, the proportion of the sample that would be host-specific herbivores, and the ratio of beetles to arthropods. Using a count of 50,000 tropical tree species, he then calculated the number of arthropod species that he would expect to find worldwide.[35] Erwin himself states that his calculations were for the purpose of creating testable hypotheses about the range of possible numbers of species actually present on Earth. Unfortunately, these speculative figures have hardened into fact in the subsequent literature.

In sum, although scientific journals and books are filled with estimates of the number of expected species on the planet, none of these estimates is very well grounded in data; at best, they are extrapolations from very limited data sets, while many are nothing more than educated guesses.

With all of these caveats in mind, let us nonetheless create a second map depicting how the taxonomic landscape would look if some of the most widely published guesses of the number of extant species were to prove true. The revised map, shown in Figure 5-14, contains several new "major players" as compared with Figure 5-13—the bacteria, fungi, algae, and viruses (if they are considered to be alive).

Figure 5-14.

A second "map" of biodiversity across the taxonomic landscape. Here we map biodiversity according to the number of species *projected* to be present on Earth (data from World Conservation Monitoring Centre, *Global Biodiversity: Status of the Earth's Living Resources* [London: Chapman & Hall, 1992], Chapter 4).

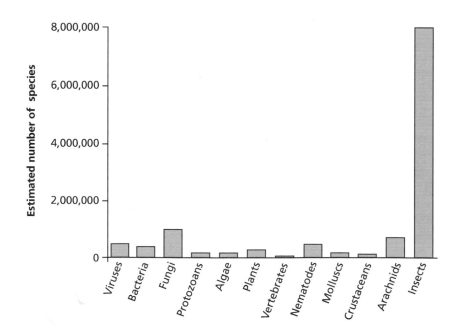

BOX
5-2.   *A brief review of the taxonomic hierarchy.*

Recall that taxonomists and systematists, biologists who study the living things of Earth in terms of their diversity and patterns of descent, have developed a seven-level hierarchy for describing the diversity of life. Each living creature has a place in this scheme, which we briefly review here. As you read this discussion, bear in mind that all of the taxonomic ranks (with the probable exception of species, a grouping that is often based on an underlying biological reality) are simply human constructs. The ranks of kingdom, phylum, class, order, family, and genus do not exist in the real world; rather, the dividing lines between them exist only in the minds of humans. Biologists have created this scheme to help organize their understanding of the diversity of life, but although the biological diversity itself is real, this scheme is nothing but human interpretation of that diversity. That said, let us review the different taxonomic ranks.

At the highest level of organization stands the kingdom. Most biologists today, as exemplified in *Five Kingdoms* by Lynn Margulis and Karlene V. Schwartz, recognize a total of five kingdoms: animals (Animalia); plants (Plantae); fungi (Fungi); bacteria and cyanobacteria (Prokaryotae); and protozoa, nucleated algae, and slime molds (Protoctista).[28] But even at this most fundamental level of distinction, biologists do not have complete consensus; some prefer to delineate eight or more kingdoms, splitting the Prokaryotae and Protoctista into multiple kingdoms.[29] Moreover, as one moves through the taxonomic hierarchy, one continually finds differences among biologists as to how to categorize the Earth's diversity. Thus, although there is often rough consensus about how to categorize the diversity of life, complete agreement never exists. Perhaps the easiest way to appreciate the organization of the taxonomic hierarchy is from the bottom up, beginning with the species level.

A Tale of Two Species

To illustrate the form and functioning of the taxonomic hierarchy, let us describe the "location" of two species within the hierarchy. *Camponotus pennsylvanicus* is the common large, black carpenter ant seen throughout the forests (and wooden homes) of the eastern United States (see Figure 5-15). *Ginkgo biloba*, the maidenhair tree, is native to East Asia but no longer exists in the wild; it is widely planted as an ornamental throughout much of the range of *C. pennsylvanicus* (see Figure 5-16). Nearly all of our students begin our class having seen carpenter ants at some point in their lives, and during the first week of school we introduce them to the maidenhair tree growing in front of our classroom building.

As you look through Table 5-2, keep in mind that all of the categories described (apart from the species) have been constructed by humans. What do the different terms in the hierarchy, such as *genus* and *family*, mean? In what sense are the families or orders of our two organisms similar or dissimilar? After all, the carpenter ant has many close relatives, many more slightly distant relatives, and so on—in fact, it belongs to one of the most densely populated regions of the taxonomic landscape. In contrast, the maidenhair tree is seen as being distinct from all other living plants. What meaning can we attach to concepts such as "diversity at the family level," when families are apparently such different entities?

Figure 5-15.

Carpenter ant. The carpenter ant of eastern North America (*Camponotus pennsylvanicus*) is found in a densely settled portion of the "taxonomic map." It belongs to a large genus, in a very large family, in one of the largest orders. Its class (Insecta) may contain almost as many species as all other classes, across the kingdoms, combined. (Photograph by Dan L. Perlman)

Figure 5-16.

Ginkgo. The maidenhair tree (*Ginkgo biloba*) is a widely planted ornamental that is taxonomically isolated. Some botanists even place it in its own division (the botanical equivalent of a phylum). (Photograph by Glenn Adelson)

Table 5-2.   A comparison of the taxonomic classification of two species.

Taxonomic Rank	*Camponotus pennsylvanicus* Carpenter ant[A]	*Ginkgo biloba* Maidenhair tree[B]
Species	*C. pennsylvanicus*	*G. biloba*
Genus	*Camponotus* several hundred described species in genus	*Ginkgo* 1 species in genus
Family	Formicidae (the ants) 297 genera and more than 8800 described species in family	Ginkgoaceae (the ginkgo) 1 genus and 1 species in family
Order	Hymenoptera (ants, bees, wasps, sawflies) Hundreds of families and more than 100,000 species in order	Ginkgoales (the ginkgo) 1 family and 1 species in order
Class	Insecta (the insects) about 30 orders and 750,000 described species in class	Ginkgoae (the ginkgo) 1 order and 1 species in class
Phylum	Arthropoda (the arthropods: insects, arachnids, crustaceans, millipedes, centipedes) perhaps 15 to 20 classes and more than 900,000 described species in phylum	Tracheophyta (the vascula plants) perhaps 10 classes and more than 280,000 described species in phylum ("phyla" are called "divisions" by botanists)
Kingdom	Animalia (the animals) 33 phyla and more than 1 million described species	Plantae (the plants) 2 phyla (divisions) and more than 300,000 described species

[A]Various authorities delineate the arthropod classes and insect orders differently, so the numbers at these ranks are approximate. Numbers of ant species and genera from Bert Hölldobler and E. O. Wilson, *The Ants* (Cambridge, Massachusetts: Harvard University Press, 1990).

[B]There is some dispute as to how to classify the ginkgo; it is worthy of its own high-level group, but just how high is under discussion. Margulis and Schwartz have placed the tree in its own phylum. The name for the order came from Michael Allaby, ed., *Concise Oxford Dictionary of Botany* (Oxford: Oxford University Press, 1992), and the name for the class from William L. Keeton and James L. Gould, *Biological Science*, 5th ed. (New York: W. W. Norton, 1993).

Cladograms: A Different Kind of Map of the Taxonomic Landscape

In Chapter 2 we described the taxonomic isolation of the tuatara, just as in this chapter we describe the taxonomic isolation of the ginkgo. In each case, we employ a different way of visualizing this isolation—two different "maps." For the ginkgo, we provide a hierarchical list with information about the number of species and number of higher taxa at each level (Table 5-2). For the tuatara, we present a phylogenetic tree, or cladogram (Figure 2-5), that graphically shows the relationship between members of the tuatara's order Sphenodontida and those of the order Squamata, consisting of the lizards and snakes. A cladogram has a particular advantage as a way of visualizing information—it uses connected branches to provide information about relationships between taxa. The elements of a hierarchical list, like sentences on a page, are generally connected in a linear series, but evolution has proceeded in a branching pattern, so a cladogram "maps" onto the underlying process that created diversity in the taxonomic landscape better than narrative description or other graphical representations.

To illustrate this proposition, consider the 15 species of the crane family (Gruidae), which includes the Eurasian crane invoked by Hesiod in the epigraph that opened this book. All of these species survive in the wild, although more than half are seriously threatened by extinction. Cranes are not only large and beautiful birds that have inspired people by their graceful majesty, spectacular calls, elaborate dances, and impressive migrations but are also specialized wetland creatures whose habitats have become threatened by development pressures throughout the world. Table 5-3 excerpted from a paper by economist Martin Weitzman, presents information about the crane family.[36]

This table provides a lot of information—more than the typical "species list" that we criticized in Chapter 4, and not just because it contains a column providing geographic range. Unlike most species lists, which admittedly have the purpose of cataloging the names of the species present in any ecological area, Table 5-3 presents important information about higher levels of taxonomy. Many species lists provide no information about the size and shape of the taxonomic units that encompass the species in those lists; that is, a list of forest trees that includes three species of oaks and four species of maples tells us nothing about the diversity of oaks in the world or about the percentage of the world's maples represented in that forest. On the other hand, Table 5-3 fills out a taxonomy; it provides every known species in each of four genera (*Balearica, Anthropoides, Bugeranus,* and *Grus*) and one family (Gruidae). But there is a corresponding incompleteness to this view. From the table we can know certain relationships within the family, for example, that members of a genus are putatively more closely related to each other than they are to the species outside that genus. But other relationships cannot be read from this list, such as the relationships among the four genera themselves and among the species within the genus *Grus*. A more information-packed method of representing the relationships within this family of birds would be to present a cladogram, as Weitzman does (see Figure 5-17).

A cladogram provides a rich historical view of relationships among species. By looking closely at Figure 5-17, we learn that the Eurasian crane is more closely related to the hooded crane than it is to the white-naped crane (the two species on either side of it in Table 5-3), information that could not have been inferred from the table alone. But there are subtleties involved in learning to read a cladogram. For instance, it is important to note that each node represents a speciation event, but each branch need not represent a species. This is true because each taxon above the rank of species (e.g., genus, subfamily, family, order, class) is thought to have originated as a single species. Thus, branches that may represent hundreds or thousands of species can be collapsed into a single branch, as has been done in the tuatara example (Figure 2-5).*

*For more detailed analyses on how to read and understand cladograms, see W. P. Maddison and D. R. Maddison, *Manual to MacClade*, version 3 (Sunderland, Massachusetts: Sinauer Associates, 1992).

Table 5-3. Crane species information list.

Common Name	Scientific Name	Geographical Range
Black-crowned crane	*Balearica pavonina*	Central Africa
Grey-crowned crane	*Balearica regulorum*	Southeast Africa
Demoiselle crane	*Anthropoides virgo*	Central Asia
Blue crane	*Anthropoides paradisea*	South Africa
Wattled crane	*Bugeranus carunculatus*	Southeast Africa
Siberian crane	*Grus leucogeranus*	Asia
Sandhill crane	*Grus canadensis*	North America
Sarus crane	*Grus antigone*	Southeast Asia
Brolga crane	*Grus rubicunda*	Australia
White-naped crane	*Grus vipio*	East Asia
Eurasian crane	*Grus grus*	Europe, Asia
Hooded crane	*Grus monachus*	East Asia
Whooping crane	*Grus americana*	North America
Black-necked crane	*Grus nigricollis*	Himalayan Asia
Red-crowned crane	*Grus japonensis*	East Asia

In this section we have provided "maps" of the crane family in table and cladogram form. In a sense, we have also provided a textual "map" of this family by providing narrative details about its ecology and conservation. None of these maps is a complete representation of reality, but all are important in understanding the context in which conservation takes place.

MAPS AND VALUES

Having read this chapter, keep the following in mind. When you look at the "real world," what you see is clearly based on what is "out there." But what you see also depends heavily on where you choose to look, including the scale at which you observe, what tools you have to help you see, and how you choose to categorize what you see—in short, your values.

Our students are able to view entire ranges of hills or plots of 1 square meter; they can use their bare eyes, hand lenses, microscopes, or laboratory instruments; finally, they can choose to assess biodiversity in terms of plant communities, taxonomic families, species, or genetic diversity. Moreover, two researchers might look at the same assemblage of biodiversity, using the same tools, selecting the same units for assessment (e.g., species), but might come up with drastically different descriptions, based on variations in how they draw boundaries around the units.

Most of the biologists studying biodiversity come from either of two traditions: systematics and ecology. When systematists view assemblages of biodiversity, they characterize them in ways very different from ecologists

Figure 5-17.

Cladogram of the crane family, Gruidae. Cladograms incorporate information about the relationships among taxa that is difficult to present in any other fashion. Compare the information content of this cladogram with that of Table 5-3. (After Martin Weitzman, "What to preserve: An Application of Diversity Theory to Crane Conservation," *Quarterly Journal of Economics* 108, No. 1 [1993]: 157–184)

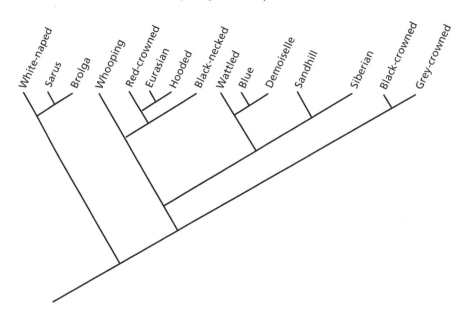

viewing the exact same assemblages. We fear from our review of the literature, that when systematists talk about assessing biodiversity, they too rarely consider the geographic landscape, and when ecologists talk about assessing biodiversity, they too rarely consider the taxonomic landscape. In short, biodiversity is present in the world, but the facets of it that you perceive and assess are constrained by your values, by where and how you look, and how you categorize what you are seeing.

References

1. Mark Monmonier, *How to Lie with Maps* (Chicago: Chicago University Press, 1991), 1.
2. See, for example, I. S. Zonnefeld and R. T. T. Forman, *Changing Landscapes: An Ecological Perspective* (New York: Springer-Verlag, 1990).
3. *Compton's Interactive Encyclopedia* (CD-ROM) (Compton's NewMedia, 1994, 1995).
4. This story comes from David Ehrenfeld, "The Management of Diversity: A Conservation Paradox," in *Ecology, Economics, Ethics: The Broken Circle,* ed. F. Herbert Bormann and Stephen R. Kellert (New Haven, Connecticut: Yale University Press, 1991), 29.
5. G. S. Hartshorn, "Plants," in *Costa Rican Natural History*, ed. Daniel H. Janzen (Chicago: University of Chicago Press, 1983).
6. World Resources Institute, *World Resources 1994–95* (Oxford: Oxford University Press, 1994), 152.
7. C. Barry Cox and Peter D. Moore, *Biogeography*, 5th ed. (Oxford: Blackwell Scientific Publications, 1993), 84.
8. T. C. Whitmore, *An Introduction to Tropical Rain Forests* (Oxford: Oxford University Press, 1990).
9. Roger Earl Latham and Robert E. Ricklefs, "Continental Comparisons of Temperate-Zone Tree Species Diversity," in *Species Diversity in Ecological Communities*, ed. Robert E. Ricklefs and Dolph Schluter (Chicago: Chicago University Press, 1993).
10. E. O. Wilson, *The Diversity of Life* (Cambridge, Massachusetts: Harvard University Press, 1992), 198.
11. Anne Magurran, *Ecological Diversity and Its Measurement* (Princeton, New Jersey: Princeton University Press, 1988).
12. D. Walker, "Diversity and Stability," in *Ecological Concepts*, ed. J. M. Cherrett (Oxford: Blackwell Scientific Publications, 1989).

13. Martin L. Cody, "Diversity, Rarity, and Conservation in Mediterranean-Climate Regions," in *Conservation Biology*, ed. Michael E. Soulé (Sunderland, Massachusetts: Sinauer, 1986).

14. Elwood C. Zimmerman, *Insects of Hawaii* (Honolulu: University of Hawaii Press, 1948).

15. E. O. Wilson, *The Diversity of Life* (Cambridge, Massachusetts: Harvard University Press, 1992), 104.

16. Axel Meyer, Thomas D. Kocher, Pereti Basasibwaki, and Allan C. Wilson, "Monophyletic Origin of Lake Victoria Cichlid Fishes Suggested by Mitochondrial DNA Sequences," *Nature* 347 (1990): 550–553.

17. E. C. Pielou, *After the Ice Age* (Chicago: University of Chicago Press, 1991).

18. M. B. Davis, "Climatic Instability, Time Lags, and Community Disequilibrium," in *Community Ecology*, ed. J. Diamond and T. J. Case (New York: Harper & Row, 1986), 269–284.

19. G. T. Prance, "Forest Refuges: Evidence from Woody Angiosperms," in *Biological Diversification in the Tropics*, ed. G. T. Prance (New York: Columbia University Press, 1982), 137–158.

20. Paul S. Martin and Richard G. Klein, eds., *Quaternary Extinctions* (Tucson: University of Arizona Press, 1984); E. O. Wilson, *The Diversity of Life* (Cambridge, Massachusetts; Harvard University Press, 1992); but see E. C. Pielou, *After the Ice Age* (Chicago: University of Chicago Press, 1991).

21. World Conservation Monitoring Centre, *Global Biodiversity: Status of the Earth's Living Resources* (London: Chapman & Hall, 1992), 196.

22. World Conservation Monitoring Centre, *Global Biodiversity: Status of the Earth's Living Resources* (London: Chapman & Hall, 1992), 196.

23. World Conservation Monitoring Centre, *Global Biodiversity: Status of the Earth's Living Resources* (London: Chapman & Hall, 1992), 197.

24. World Conservation Monitoring Centre, *Global Biodiversity: Status of the Earth's Living Resources* (London: Chapman & Hall, 1992), 197.

25. World Conservation Monitoring Centre, *Global Biodiversity: Status of the Earth's Living Resources* (London: Chapman & Hall, 1992), 197.

26. Note, however, that the distribution of genetic diversity across the taxonomic landscape is still little researched. Moreover, to the extent that genetic diversity *has* been studied, the results are often applied to mapping the taxonomic landscape, so we have a problem of circularity. We cannot both use genetic diversity to create our map of the taxonomic landscape and ask how it is distributed through the landscape.

27. These numbers were calculated from information in E. O. Wilson, *The Diversity of Life* (Cambridge, Massachusetts: Harvard University Press, 1992).

28. Lynn Margulis and Karlene V. Schwartz, *Five Kingdoms*, 2d ed. (New York: W. H. Freeman, 1988).

29. William T. Keeton and James L. Gould, *Biological Science*, 5th ed. (New York: W. W. Norton, 1993).

30. World Conservation Monitoring Centre, *Global Biodiversity: Status of the Earth's Living Resources* (London: Chapman & Hall, 1992), 36.

31. World Conservation Monitoring Centre, *Global Biodiversity: Status of the Earth's Living Resources* (London: Chapman & Hall, 1992).

32. World Conservation Monitoring Centre, *Global Biodiversity: Status of the Earth's Living Resources* (London: Chapman & Hall, 1992), 19. The number of 80,000 species of nematodes comes from Lynn Margulis and Karlene V. Schwartz, *Five Kingdoms*, 2d ed. (New York: W. H. Freeman, 1988).

33. Terry L. Erwin, "Tropical Forests: Their Richness in Coleoptera and Other Arthropod Species," *The Coleopterists Bulletin* 36 (1982): 74–75; Nigel E. Stork, "Insect Diversity: Facts, Fiction and Speculation," *Biological Journal of the Linnean Society* 35 (1988): 321–337.

34. Terry L. Erwin, "How Many Species Are There?: Revisited," *Conservation Biology* 5 (1991): 330–333.

35. Terry L. Erwin, "Tropical Forests: Their Richness in Coleoptera and Other Arthropod Species," *The Coleopterists Bulletin* 36 (1982): 74–75.

36. Martin Weitzman, "What to Preserve: An Application of Diversity Theory to Crane Conservation," *Quarterly Journal of Economics* 108, no. 1 (1993): 157–184.

6

Ambiguities

THE BOUNDARY AMBIGUITY

During the first week of the fall semester we take our students to Wompatuck State Forest, south of Boston, Massachusetts, in order to start them thinking about biodiversity. We ask them to determine on their own, without help from instructors, handbooks, or species lists, how many different kinds of trees they can find in the forest. When we use the phrase "different kinds of," we have a hidden agenda—to have our students begin to wrestle with the identification of *species*. Students have little trouble noticing that all white pines (*Pinus strobus*) are similar to each other and are different from any other type of tree, although when pitch pine (*Pinus rigida*) is present, they can see similarities between white pines and pitch pines that distinguish them from all other non-pine trees in the forest. Still, students readily can distinguish between pitch pine and white pine on the basis of several obvious characteristics: pitch pine always has three needles in a group, and white pine always has five; pitch pine has a stubby cone with a sharp tip to each scale, while white pine has an elongate cone with smooth scales.

The forest also contains red oak (*Quercus rubra*), black oak (*Q. velutina*), and scarlet oak (*Q. coccinea*), three species that are far more difficult to differentiate with an untrained eye. The leaves, buds, fruits, bark, and overall tree shape are quite similar among these three oaks (see Figure 6-1), and even a trained biologist can have trouble distinguishing them solely by certain of these characteristics, such as leaf shape. Not surprisingly, several students fail to identify these three oaks as separate species. However, once the students have been shown the distinguishing characteristics (e.g., size, shape, and markings on the acorn), they quickly are able to identify many of the oaks in the forest to species—but, curiously, not all.

A fundamental difficulty in identifying individuals of these three species of oaks goes far beyond the problem of learning how to see technical characteristics—the red, black, and scarlet oaks hybridize with one another and produce offspring that possess characteristics that are midway between those listed in field guides as representative of each of the parent species. For example, field guides tell us that the black oak's buds are sharply angled and densely hairy, while the red's are less angled and sparsely hairy; that the bark of the black oak has dull stripes that are broken up into alligator-skin-like blocks near the base of the trunk, while the red oak's bark is furrowed with long, shiny stripes. But try gathering a dozen students around a 20-meter-tall enigma whose buds are sharply angled and sparsely hairy, and whose furrowed bark is dull yet not broken into blocks at the base. When they first gaze upon a hybrid

Figure 6-1.
Three species of oaks. (A) The red oak, (B) black oak, and (C) scarlet oak are difficult to distinguish from each other using vegetative characters alone, as these drawings of leaves and buds illustrate.

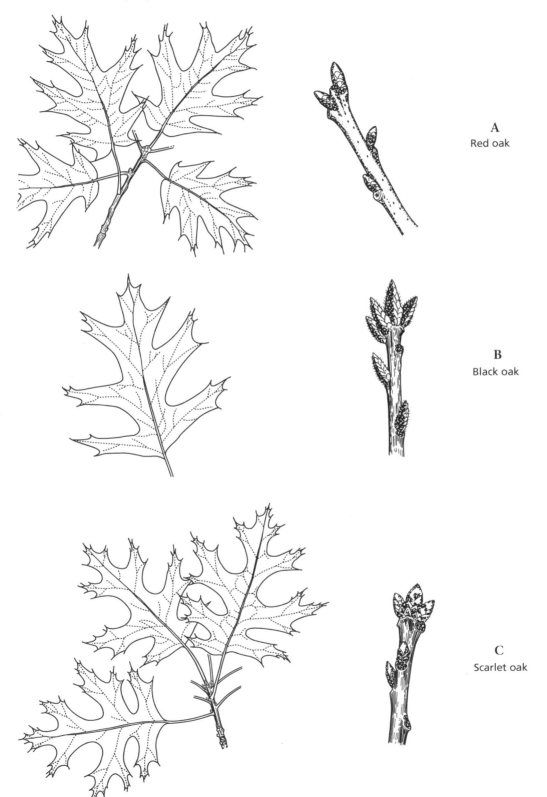

A
Red oak

B
Black oak

C
Scarlet oak

oak and begin to contemplate questions like "Should red oak and black oak be kept as separate species or should they really be considered a single species?" their world views begin to change. Most students come into our class concerned about species extinction and preservation of endangered species, without any sense that the definition of species is a problematic issue; they soon gain an understanding of the difficulty of drawing unambiguous boundaries between those chunks of nature we call "species."

The students' confusion heightens even more when, during the same weekend, we visit Ames Pond on the island of Martha's Vineyard (see Figure 6-2). As we examine the pond with our students and begin to observe carefully the life-forms in and around it, the first question we ask is, "How would you define the boundaries of the pond ecosystem?" Some of the students recall from their ecology classes that an ecosystem is a bounded system consisting of a biotic community and its physical environment, with relatively little energy or nutrient flow across the ecosystem's borders. With this definition in mind, someone invariably suggests the water's edge as the boundary for the pond ecosystem. Within moments, however, another student will point to the stream flowing out of the pond, and the trees overhanging the water—both of which contribute greatly to energy and nutrient flow across the boundary of the pond.

Soon, after scooping dozens of aquatic insects from the pond, the students find that most of these organisms are not full-time residents of the pond. Many are larvae or nymphs that will emerge and fly away from the pond later in life, while others are already winged adults and can leave the pond at any time. If we are lucky, we might see a kingfisher dive into the pond and permanently remove one of the fish that, until that moment, seemed a nontransient member of the pond ecosystem. With the boundaries of the biotic community so porous, we begin to see that delineation of the ecosystem itself is a difficult task.*

The oaks and the pond both illustrate the boundary ambiguity, one of many ambiguities we run into in our attempts to describe the natural world and assess biodiversity. In theory, every species and every ecosystem is

---

*A brief clarification of usage of the terms *ecosystem* and *community* is necessary here. The two terms have always been intimately related; an ecosystem is typically viewed as a biological community interacting with its physical environment, as the following definitions from *The Cambridge Illustrated Dictionary of Natural History* illustrate:

> *Community:* Any group of organisms comprising a number of different species that co-occur in the same habitat or area and interact through trophic and spatial relationships.

> *Ecosystem:* A community of organisms and their physical environment interacting as an ecological unit.

In speaking of biodiversity, it may in fact be logically more proper to speak of communities rather than ecosystems; for example, the groups of trees at Wompatuck State Forest that we discuss in the opening paragraphs of this chapter are often referred to as the oak-pine *community*. However, most of the conservation and biodiversity literature uses the term *ecosystem* when discussing interacting groups of individuals localized in space, as evidenced by definitions of biodiversity that discuss variety in species, genes, and ecosystems. For the purpose of our analyses in this chapter, we use the term *ecosystem* as it is used in the biodiversity literature—to encompass the meanings of both *ecosystem* and *community* as used by community ecologists.

Figure 6-2.
The Ames Pond ecosystem. A pond seems like one of the easier ecosystems around which to draw a clear boundary. But which elements of biodiversity are really inside and outside this ecosystem? (Photograph by Armando Carbonell)

"bounded." That is, for any given item in the world, one should be able to say whether or not it is a part of the entity called "the red oak species" or the entity called "the Ames Pond ecosystem." Many species have reasonably good boundaries, although many do not. An individual tree is either a ginkgo or it is not; the boundary between ginkgo and not-ginkgo is quite clear. As we saw with the oaks, however, some species have fuzzy boundaries, and later in the chapter we discuss other species that exhibit different types of fuzzy boundaries. Ecosystems, in contrast, all have fuzzy boundaries. No ecosystem is so clearly demarcated that it can be said to exist without outside influences; the geographic landscape contains no unambiguous boundaries. But *species* and *ecosystem* are key terms in the conceptualization of biodiversity and the applied science of how to conserve it. Should we worry about the boundary ambiguity and the problem of fuzzy boundaries?

The answer to this question is that there are times to worry and times not to worry. Consider the gray wolf (*Canis lupus*), a species that the U.S. government and many conservation organizations have put a significant amount of effort and money into protecting, and even reintroducing to its former range (see Figure 6-3). The gray wolf is known to hybridize with the domestic dog (*Canis familiaris*). However, this fact is of little practical concern, as the hybrids are easily recognized, and it is unlikely that effort and money will be wasted preserving wolf-dog hybrids. On the other hand, the endangered red wolf (*Canis rufus*) is believed by some investigators to

Figure 6-3.
North American species of the genus *Canis*. The gray wolf (*Canis lupus*), the red wolf (*Canis rufus*), the coyote (*Canis latrans*), and the domestic dog (*Canis familiaris*) are four species in the same genus. Hybridization has been documented between all pairs of these species. Some of these hybridizations are problematic for conservation and some are not. (Drawings not to scale)

Gray wolf

Coyote

Red wolf

German shepard

have initially arisen from either an ancient or recent hybridization between gray wolves and coyotes (*Canis latrans*). One of the U.S. government's first successful captive breeding programs under the Endangered Species Act involved the conservation of the red wolf. However, according to a later policy decision by the U.S. Fish and Wildlife Service, hybrids became ineligible for protection under the Endangered Species Act. Did we waste valuable conservation effort and resources protecting what was essentially a mistake due to a leaky boundary between two species? Some commentators say "yes" and others "no"—either way, understanding the boundaries among the species of the genus *Canis* is essential to making further decisions about the conservation of the red wolf.[1]

In this chapter we highlight a number of ambiguities that arise in the key terms that have been used to define biodiversity—*species, ecosystem,* and *gene*. We have already devoted several of the early chapters to discussions of ambiguities inherent in the terms *diversity*, *biodiversity*, and *values*. By discussing the ambiguities of all these terms in detail, we aim to help conservation decision makers understand the ways in which assessments of biodiversity are consistent and rigorous, and those in which they are ambiguous. The terms that we discuss contain ambiguities that fall into two categories: those that are ambiguous in nature itself (as with the oaks and the pond) and those that are ambiguous in the ways that humans speak and think about nature. Some terms are ambiguous in both ways.

THE BOUNDARY AMBIGUITY IN THE TEMPORAL LANDSCAPE: THE PROBLEM OF EVOLVING ENTITIES

Boundaries in the temporal landscape can be as ambiguous as they are in the geographic and taxonomic landscapes. This problem is especially acute for entities that evolve or change their character over time, such as species and ecosystems. The evolution of species in the genus *Homo* (*H. habilis*, *H. erectus*, and *H. sapiens*) is an example of this type of ambiguity. Although there is disagreement about the delineation of these three species, many anthropologists agree that some *H. habilis* populations evolved into *H. erectus*, some populations of which evolved into *H. sapiens* (see Figure 6-4).[2] However, this evolution was a gradual process, not a series of discontinuous leaps. No boundary existed in time before which all earlier individuals belonged to one species, say *H. erectus*, and all later individuals belonged to *H. sapiens*.

An ecosystem undergoing succession also creates boundary ambiguities for the conservation biologist attempting to place it in a clearly defined category. While one can find ecosystems that are readily characterized as old fields, and others that are early successional forests or late successional forests, these categories do not have clear dividing lines between them. Over time, old fields slowly fade into young forests, which grow into older forests. In contrast, events that reset succession, such as catastrophic fires or windstorms, are generally easy to locate in time. What was an old forest yesterday is today merely an area of burned or uprooted trees about to become a young forest again.

The conservationist attempting to categorize the world's biodiversity will often run into the problem of ambiguous boundaries in time and be unable to assign an element of biodiversity to a single species or ecosystem type. Evolution and succession are slow processes relative to the time scales on which humans perceive the world. The differences between the scale of our perception and the rate of some natural processes cause what we call the ambiguity of evolving entities, which can confuse our attempts to understand and protect biodiversity.

We have already presented two kinds of ambiguity, the boundary ambiguity in space and the boundary ambiguity in time, that are inherently

Figure 6-4.
Hominid cladogram. The phylogenetic relationship between some of the species of the genus *Australopithecus* and the three species of the genus *Homo*. Was there a clear boundary between any of the species? (After Richard Cowen, *History of Life*, 2nd ed. [Boston: Blackwell Science, 1995])

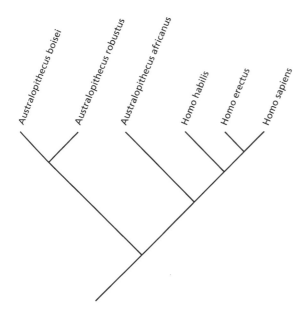

ambiguous in nature. We now turn to a series of three ambiguities that stem largely from the ways that humans speak and think about nature:

- the ambiguity of scale
- the difficulty of distinguishing a parcel from a type
- the ambiguity of the phrase "genetic diversity"

Although these are qualitatively different from ambiguities inherent in nature, they are no less problematic for conservation decision makers attempting to assess biodiversity.

THE AMBIGUITY OF SCALE

The term *ecosystem* refers to entities that can be described and delineated at many different scales, which makes the term quite ambiguous. Ames Pond, for example, lies wholly within the Cedar Tree Neck Preserve, an ecosystem that includes other pond ecosystems, bog ecosystems, and a few different types of forest ecosystems. The conservation decision maker can think of ecosystems at the scale of Ames Pond or Cedar Tree Neck—both fit standard definitions for ecosystems. Alternatively, the decision maker might wish to consider a still larger ecosystem, such as the entire island of Martha's Vineyard, or a smaller ecosystem, such as the gut inhabitants of a single dragonfly nymph in Ames Pond (see Figure 6-5).

The term *ecosystem* reflects an ambiguity of scale because the same term is used, according to plant ecologists Fakhri Bazzaz and Tim Sipe, for any level "in the hierarchy of open living systems, from cell to planet, that transforms energy and resources."[3] Because of the ambiguity of scale, conservation decision makers cannot readily use ecosystems as units for comparative purposes because the ecosystems being compared are frequently defined at different scales. Just as it is not possible to count the number of cookies that can be made from a certain batch of dough—cookies can be made at any scale—so, too, the ecosystems of a given landscape cannot be counted.

Figure 6-5.
Nested ecosystems at Cedar Tree Neck Reserve. The term *ecosystem* is ambiguous because the same term applies at every scale. Ames Pond is an ecosystem within the surrounding beech-maple forest, which is an ecosystem within the surrounding Cedar Tree Neck Reserve, which is an ecosystem within the surrounding island of Martha's Vineyard, which is also an ecosystem.

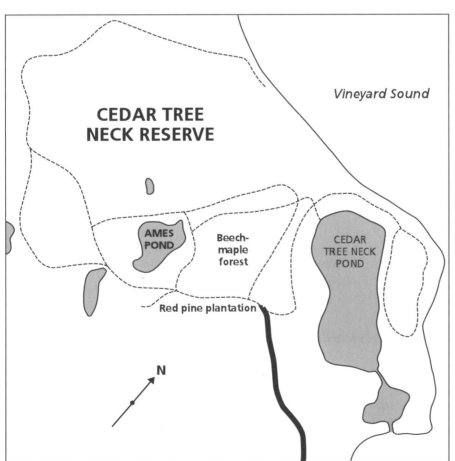

Ecosystems can be nested within each other, up to the level of the entire planet, and no single scale exists for counting or otherwise studying ecosystems. Thus, decision makers must specify the scale of ecosystems they are considering.

IS AN ECOSYSTEM A PARCEL OR A TYPE?

As a part of attempting to "protect biodiversity," conservation decision makers attempt to "protect ecosystems." At first glance this seems like a reasonable task, but upon reflection, how does one go about protecting ecosystems? What exactly is one supposed to protect? We have already found that the term *ecosystem* is ambiguous because of leaky boundaries and problems of scale. In this section we discuss a further ambiguity: the confusion between ecosystems as physical entities (e.g., a parcel of land that one could put a fence around) and ecosystems as abstract types. Consider the following description of the term *ecosystem* by the noted ecological theorist Robert May:

> The ecosystem may be a particular community (e.g. Harper's grass-field at Aber in North Wales); it may be some environmental type (e.g. the mangrove ecosystem), or a geographic region (e.g. the Amazonian ecosystem), or even the whole world (the global ecosystem).[4]

A careful reading of May's passage shows this ambiguity at work. The first ecosystem that he mentions, Harper's grassfield, is a particular physical entity, what we will call a *parcel* of terrain. But he quickly moves on to note that the term *ecosystem* may also stand for an "environmental type," which, like the mangrove ecosystem, includes many parcels (patches of terrain) that share certain similarities. He also includes the "geographic region" and "the whole world," both of which can be viewed as simply very large parcels (see Figure 6-6). It is essential that conservation decision makers specify which of these different sorts of "ecosystems" they are targeting when they attempt to protect ecosystems. Not only are the reasons for protecting parcels and types quite different, but the two categories require different strategies for protection.

A parcel might be proposed for protection for any of several reasons: it might be a particularly beautiful piece of land, it might perform some important ecosystem service, it might contain organisms found nowhere else, or it might contain a unique assemblage of species. Moreover, it is relatively clear what one might do to protect a parcel; one could purchase (or secure conservation easements for) the land along with an appropriate buffer around the site. Following acquisition, one would develop a management plan to help protect the site from unwanted influences such as exotic species, excessive nutrient flow into the site, or certain successional patterns (e.g., the development of forest at a site where one aims to retain an open field).

As part of protecting biodiversity, one could make a case for protecting ecosystem types. How, though, would one protect "the mangrove ecosystem," an ecosystem type spread along the coasts of five continents? An ecosystem type, as an abstract entity, cannot itself be protected. We can only protect examples of types, namely, parcels. Thus, another reason for protect-

Figure 6-6.

Different "kinds" of ecosystems. The term *ecosystem* is ambiguous because it is sometimes used to refer to a particular parcel and other times used to refer to an abstract type. Harper's grassfield, the mangrove ecosystem, the Amazonian ecosystem, and the whole world are all examples of *ecosystems*, but are they examples of the same kind of thing?

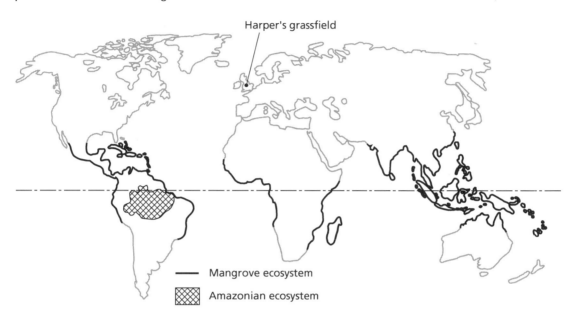

ing a specific parcel might be that it serves as an example of an ecosystem type, such as mangrove swamps or boreal forest. The parcel that May calls "Harper's grassfield" plays an important role as one example of a type of ecosystem, namely, grassfields. Presumably, Harper did not study his grassfield because it was unique; rather he aimed to discover lessons and principles that could be applied to similar grassfields elsewhere. Two key issues concerning the protection of ecosystem types remain, however. First, one needs to select a specific classification scheme to determine which ecosystem types to protect. The classification scheme used for planning the protection program has a tremendous effect; if one selects a classification system that focuses only on gross differences between ecosystem types and defines only a few types, then one need only protect a few parcels. If, instead, the system emphasizes slight differences between ecosystem types and delineates large numbers of ecosystem types, then it will require the protection of a huge number of examples[5] (see Figure 6-7). Second, conservation decision makers need to develop clear criteria for knowing when the job is complete and the ecosystem type is "protected," a process that will depend heavily on the values and goals of the decision makers.

According to our reading of the literature, ecologists tend to use the term *ecosystem* to refer to parcels, while conservationists often use the term to refer to types, as shown by the following comparison. In a 1993 report for the Congressional Research Service, M. Lynne Corn clearly treats the entities she calls "ecosystems" as parcels:

Figure 6-7.

Classifying ecosystem types. The figure shows (A) the seven major habitat types and (B) the 27 regional habitat units drawn up by the Biodiversity Support Program for South America. Are there seven ecosystem types that need to be included in the conservation strategy? Are there 27 ecosystem types that need to be included? Or are there more than 27? See Chapter 8 for more on the decision-making process of the Biodiversity Support Program. (Biodiversity Support Program, Conservation International, The Nature Conservancy, Wildlife Conservation Society, World Resources Institute, and World Wildlife Fund, *A Regional Analysis of Geographic Priorities for Biodiversity Conservation in Latin America and the Caribbean* [Washington, D.C.: Biodiversity Support Program, 1995], 3)

A

Tropical Moist Lowland Forests

Tropical Moist Montane Forests

Tropical Dry Forests

Xeric Formations

Herbaceous lowland Grasslands

Herbaceous Montane Grasslands

Temperate Forests

Figure 6-7.
(continued)

Tropical Moist Lowland Forests

1-1   Atlantic
1-2   Upper Amazon
1-3   NE Amazon
1-4   SE Amazon
1-5   Chocó-Darién

Tropical Moist Montane Forests

2-1   Tropical Andes Montane Forest
2-2   Venezuelan Coastal Montane Forest
2-3   Guayana Montane Forest

Tropical Dry Forests

3-1   N South American Dry Forest
3-2   Western Andes Dry Forest
3-3   Chaco
3-4   Cerrado-Pantanal

Xeric Formations

4-1   Caribbean Xerics
4-2   Caatinga
4-3   Peru-Chile Deserts
4-4   Chilean Winter Rainfall Xerics
4-5   Argentine Monte

Herbaceous Lowland Grasslands

5-1   Llanos-Gran Sabana
5-2   Pampas
5-3   Patagonian Steppe
5-4   Amazonian Savannas

Herbaceous Montane Grasslands

6-1   Paramo
6-2   Puna
6-3   Southern Andean Alpine
6-4   Pantepui

Temperate Forests

7-1   Southern Temperate
7-2   Brazilian *Araucaria*

The word [ecosystem] is rarely, if ever, applied to disjunct spaces: two similar mountaintops would ordinarily be considered *similar* ecosystems rather than *an* ecosystem, unless the land between them is also included. Instead, the word *biome* is usually applied to visually similar but not necessarily connected areas.[6]

In contrast, the book *Global Biodiversity* contains a color map entitled "Major World Ecosystems." This map of the entire planet depicts 46 "ecosystems" in different colors and patterns, including tundra, hot desert, tropical savanna, mangroves, and cool conifer—each of which appears on multiple continents, and each of which clearly refers to ecosystem types. The regions of tundra in central Canada and Siberia, which are shown as part of a single ecosystem on the map, are clearly disjunct; Corn would therefore call them "similar ecosystems." Is there a good reason for this distinction in usage between the two disciplines, or is it merely sloppiness? To answer that question, we need to return to another important concept of community ecology—that of succession.

PROBLEMS WITH PROTECTING PARCELS

To illustrate this concept, let us take a closer look at the parcel known as the Hyannis Ponds ecosystem. These ponds, which form the basis of the class role-play with which we opened Chapter 3, are examples of an ecosystem type known as the coastal plain pond. According to a report by the Massachusetts Natural Heritage and Endangered Species Program, Mary Dunn Pond, one of the Hyannis Ponds, is the finest remaining example of the coastal plain pond shore ecosystem left in New England. By dint of that status, this pond may be the finest example of the coastal plain pond shore ecosystem left in the world, as the ecosystem type appears only along the eastern seaboard of North America.* Throughout the summer, the shores of Mary Dunn Pond are populated by various wildflowers that are not commonly found in other habitats in the eastern United States. In late July and early August, there is generally a profusion of Plymouth gentian (*Sabatia kennedyana*), a flower of 8 to 12 petals, bright rose-pink with a yellow circle etched into its center. One can also find a narrow-leaved species of arrowleaf (*Sagittaria teres*), a thread-leaved species of the carnivorous sundew (*Drosera filiformis*), and many other plant species that are found only in coastal plain

---

*The concept of ecosystem as *type* is also subject to the boundary ambiguity, as the discussion of the problem of classification schemes for ecosystem types highlights. As an example of this ambiguity, how can we evaluate the claim that Mary Dunn Pond is the best example of the coastal plain pond shore ecosystem in the world? How does it compare to a coastal plain pond shore in North Carolina that has a 60% overlap in community-specific plant species? If there is only a 60% overlap in community-specific plant species, then should the two physical entities, that is, the pond shore in North Carolina and the pond shore in Massachusetts, be part of the same ecosystem type? There is a fuzzy boundary here, one that might allow us to separate the North Carolina coastal plain pond shore as a different ecosystem type, perhaps the southern coastal plain pond shore, to be distinguished from the northern coastal plain pond shore of Mary Dunn.

Figure 6-8.
Pine seedlings invading Little Israel Pond. A coastal plain pond ecosystem, if it dries out, will turn into an oak-pitch pine ecosystem. Thus, the same particular parcel of land can be classified as different types of ecosystems at different times. (Photograph by Dan L. Perlman)

pond shore ecosystems (see Color Plates 12, 13, and 14). On sunny summer days, dragonflies and damselflies flit about the pond's edge, and if one is lucky, one might see the long-legged green darner (*Anax longipes*), one of the largest dragonflies found in the United States. The pond is shallow, so much so that in the particularly dry summer of 1991, there was no water at the surface, and visitors could walk across the cracked clay of the pond's bottom without getting their feet wet. Surrounding Mary Dunn Pond and the other Hyannis Ponds is an oak–pitch pine ecosystem, a native ecosystem that is far and away the commonest forested ecosystem type on the Cape.

About a third of a mile north of Mary Dunn Pond lies a geological depression known locally as "the field" or "the meadow" or "Little Israel Pond." This area was formerly a coastal plain pond, but over the years the standing water, which is separated from the water table by a clay pan, has dried out more or less permanently and only returns during the few weeks after a heavy rain. The entire depression is now covered with an herbaceous plant community that includes the Plymouth gentian, and it is there that nature lovers can get a spectacular view of the beautiful gentian in copious flower—for nowhere in New England is there as large a population. However, an observant visitor might note that scattered throughout the magenta-filled show of gentian flowers that appear each July are numerous pitch pine seedlings and saplings, patiently changing what was once a coastal plain pond, and is now essentially a coastal plain pond shore community (without the pond), into a future oak and pitch pine forest (see Figure 6-8). In Mary Dunn Pond and the other water-filled ponds of the Hyannis Ponds system, periodic inundation by high water kills invading pine seedlings and maintains the local community of wildflowers and insects characteristic only of the coastal plain pond shore. However, Little Israel Pond is undergoing a conversion, and unless several years of high water come back to the meadow to drown the young pines, in 30 or 50 years the meadow will become much like the rest of Cape Cod, an example of the oak–pitch pine ecosystem type.

Thus, the natural dynamics of the ecosystem can eliminate the very species and community that conservation decision makers are attempting to conserve. To protect the Plymouth gentian and coastal plain pond shore community at Mary Dunn Pond, it is not enough to buy and set aside a parcel of land. The conservation decision maker must understand the ecology of the community and the forces of change at work within the ecosystem—and must be able to act on this knowledge. Furthermore, the decision maker must understand the time scale on which these forces act.

In conservation biology, protection plans are often created on a scale of several human generations or centuries. For example, Mark Shaffer and others who study the long-term viability of populations have defined a minimum viable population as the smallest isolated population having a 99% chance of surviving for 1000 years.[7] Conservation decision makers, however, are often called upon to respond to threats that operate on a much shorter time scale, often as short as 1 to 10 years in the cases of ecosystems recently exposed to dramatic disturbance, human development pressures, or invading foreign species. Moreover, succession can have a major impact on a parcel of land on the time scale of 10 to 100 years—well within the time frame in which conservation decision makers aim to protect ecosystems. When one categorizes parcels by type, calling one parcel an example of the coastal plain pond ecosystem and another an example of the oak–pitch pine ecosystem, one finds that within the lifetime of an observer a particular parcel may change from an example of one type of ecosystem to an example of another type. Today's coastal plain pond shore community of Little Israel Pond may be an oak–pitch pine forest 50 years from now.

Understanding succession is crucial to understanding the different meanings of the term *ecosystem*. Conservationists are in for rude shocks when they rely on protecting ecosystems as parcels only, that is, if they put a fence, literally or figuratively, around a site and expect it to remain unchanged. Such a surprise occurred in one of the world's smallest nature preserves, the Badgeworth Nature Reserve in Gloucestershire, England, which was established to protect a rare marsh buttercup, *Ranunculus ophioglossoides*. The reserve was officially constructed by the placement of a barbed-wire fence around a marsh that was home to a large population of the marsh buttercup. In subsequent years, the area inside the fence was often found to contain no buttercups at all, while the land outside the fence was often covered with hundreds of these flowers. It turned out that grazing and trampling by cattle was necessary to prevent the succession to a taller plant community that shaded out the buttercups.[8]

For community ecologists, who study the mechanisms that regulate change in ecosystems or change in species diversity within an ecosystem, a parcel that changes from being an example of one ecosystem type to another is a data point, something that provides vital information about the mechanisms they are trying to understand. For conservationists, the succession of a protected parcel from being an example of a coastal plain pond to an example of an oak–pitch pine forest is a loss.

For this reason, conservationists cannot afford to look at ecosystems only as parcels; they must also abstract to type. Only by understanding the dynamics of succession and the spatial and temporal distribution of ecosystem types can one intelligently plan to "protect ecosystems" in a way that will survive the loss of individual parcels when they change from examples of one type of ecosystem to examples of another type through succession. In other words, the particulars of successional patterns of ecosystems and the context of a particular example of an ecosystem relative to other examples of its type are the key pieces of information that a conservationist needs to conceptualize how to "protect ecosystems." The type/parcel ambiguity does not resolve itself into a battle in which one of the two is declared a winner. Conservation decision makers need to conceptualize ecosystems both as types and as parcels in order to best understand how to conserve them. The important thing is to distinguish when one is referring to the parcel and when one is referring to the type, and why.

"GENETIC DIVERSITY": A TERMINOLOGICAL AMBIGUITY

Biodiversity is usually defined in terms of genes, species, and ecosystems. Yet rarely do we see a conservation plan that focuses on genes as the primary element for protection. We have never seen a discussion of the confusion between the two different things that the phrase "genetic diversity" has come to mean in the world of biodiversity conservation. First, specific genes are elements of biodiversity to be protected for their worth in specific situations. For example, genes in wild relatives of corn or rice may be important in conferring resistance to a certain disease, or a gene in a plant may code for a powerful antiviral agent that can save human lives. These genes can be conserved in living natural populations, cultivated populations, or gene banks; they can even be isolated and sequenced, and preserved as information in a computer. Such genes can be the subject of property rights, and some of the most contentious issues in the negotiations leading up to the Convention on Biological Diversity were on this subject. Clearly, the greater the diversity of such genes that our species has access to, the more promising our future looks. The future holds many surprises for the health of humans and of the domesticated species upon which we depend for food; the greater the diversity of genetic responses that we can call upon, the better off we will be. Almost without exception, this form of genetic diversity is discussed in the context of species or varieties that are directly useful to humans, rather than wild species being helped to survive for their own sakes.

The second form of genetic diversity is generally agreed to be important for the survival of a population—it is not any specific gene that is important, it is genetic diversity itself. Maintenance of genetic diversity in this form prevents inbreeding depression and the accumulation of lethal recessive traits in small populations. This concept has long been the central focus of the biological subdiscipline of population genetics, in which it is called genetic variation, and has been expressed quantitatively in many different measures. Papers that discuss the measurement of biodiversity in terms of genetics, such as found in Kevin Gaston's book, *Biodiversity: A Biology of Numbers and Difference*, concern themselves with this aspect of genetic diversity.[9] Such expressions of genetic variation are necessarily summary statistics and do not represent any real thing in the world, anything that can be isolated, sequenced, saved, or subjected to property rights. This form of genetic diversity is typically discussed in the context of wild populations.

INVALID COMPARISONS AND THE INCOMMENSURABILITY OF SPECIES

A problem that has long plagued evolutionary biologists is how to draw boundaries between species. This problem appears as the boundary ambiguity in morphology and breeding that we illustrated at the beginning of this chapter with the example of the oaks. It also includes the boundary ambiguity in time or the ambiguity of evolving entities, namely, how to decide when one species has evolved into another, that we illustrated with hominid evolution.

However, for the conservation decision maker, there is a third facet to this problem. This results from our using a single term, *species*, to describe entities that are defined and bounded in very different ways—ways that cannot truly be compared, ways that are incommensurable. This incommensurability stems in part from the variety of evolutionary factors at play during the origin and maintenance of different species, in part from the way that biologists partition the natural world into species. Even were we perfectly able to assign every individual organism and every population unambiguously to a species, it still does not mean that the different types of species would be comparable in any meaningful way, that they could be added together as if they were the same kind of units. The taxonomic and temporal boundary problems we discussed earlier involved boundaries that were sufficiently unclear that it would be difficult to place individuals unambiguously on one side of the boundary or the other. The essence of the incommensurability ambiguity is that we are considering a number of different boundaries, typically boundaries between species, and we find that the species themselves are not all the same class of items and that they cannot consistently be compared with each other.

The issue of how to define species has been the subject of an exhausting literature and is still one of the most contentious issues in evolutionary biology.[10] Our purpose here is not to provide a comprehensive review of the "species problem," but to provide the reader with enough background to understand how the concept of species is beset by the ambiguities of leaky boundaries, evolving entities, and incommensurability, and why the assessment of biodiversity must take these ambiguities, especially the last, into account.

Most basic biology textbooks provide students with some variation of Ernst Mayr's classic definition of species, known as the *biological species concept:* "Species are groups of actually or potentially interbreeding natural populations, which are reproductively isolated from other such groups."[11] Most practicing biologists, if asked for their species concept, would reply with something like Mayr's definition. If they responded differently, they would probably choose another theoretical concept, such as the *evolutionary species concept* (a single lineage of ancestor/descendant populations of organisms that maintains its identity from other such lineages and that has its own evolutionary tendencies and historical fate) or the *phylogenetic species concept* (the smallest diagnosable cluster of individual organisms within which there is a parental pattern of ancestry and descent).[12]

However, none of these theoretical species concepts has been used to circumscribe the vast majority of species that have been described in the literature, since these concepts are virtually impossible to employ in practice. Instead, the species concept actually used to delimit most species has generally been a practical, morphological concept, based on differences in physical characteristics among specimens (with occasional behavioral or genetic characteristics also used for delimiting species).

A characteristic that all species concepts have in common, however, is that they necessarily assume that species are formed and maintained in a uniform manner; that is, that all species are the same kind of thing. Yet we know, for example, that species are not all formed in the same manner and are not the same kind of thing. An obvious and often-used example is the phenomenon of asexual organisms. All monerans (i.e., bacteria and cyanobacteria), many protists (mostly single-celled organisms with a true nucleus), and even some higher plants and animals are asexual. Hundreds, and perhaps thousands, of plant "species" are obligately asexual, as are several "species" of lizards and salamanders.[13] And bacteria, although they can exchange genetic material, do not have the meiotic apparatus to be considered truly sexual. None of these asexual "species" is tractable to the biological species concept, because not a single organism or population can interbreed with anything else. Interbreeding is a concept that has no meaning when applied to these asexual taxa.

Consider the ramifications of this phenomenon when it comes to the matter of comparing species across the vast taxonomic landscape. How do we determine what a bacterial species is? By stable differences in morphological and genetic characteristics. What caused those differences? How different is one bacterial species from its closest relative? What does it mean to be a closest relative? Does it mean something different in bacteria from what it means in higher animals and plants? In any case, we are talking about two very different kinds of species when we talk about bacterial species and, for example, bird species.

In addition to asexuality, hybridization causes commensurability problems for species. Hybridization is a common biological phenomenon throughout all sexually reproducing groups. It is most well-known in plants, but occurs throughout the animal kingdom, and is even reasonably common in vertebrates.[14] Botanist Lucinda McDade notes that the term *hybridization* itself is subject to multiple meanings, but for our purposes, we can narrow the meaning to sexual reproduction between individuals of two different species that produces fertile offspring. Hybridization may be evidence of incomplete speciation between the two parent species; that is, because the

Figure 6-9.
Cladogram of the genus *Sabatia*. Only *Sabatia kennedyana* and *S. campanulata* are noted; all other branches represent other species of *Sabatia* that are not mentioned in the text. *S. kennedyana* and *S. campanulata* are known to hybridize even though they are not closely related to one another within the genus *Sabatia*.

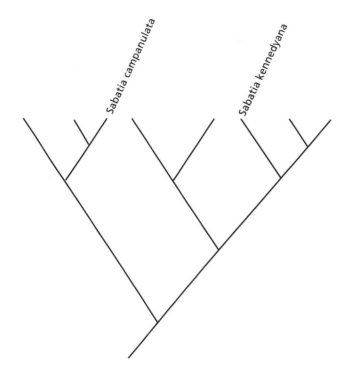

speciation process may take thousands of years to occur, at any given moment on Earth there are many species in the process of splitting into two or more new ones. This is one possible explanation for the hybridization among the red, black, and scarlet oaks that we mentioned at the beginning of this chapter. But hybridization is not necessarily evidence of incomplete speciation—it may occur between two species that are not closely related to one another within a genus, such as the hybridization between the Plymouth gentian (*Sabatia kennedyana*) and the marsh pink (*Sabatia campanulata*) that has been verified at the Hyannis Ponds.[15] These two *Sabatias* are in different subgenera of the genus, so that at least three complete speciation events have taken place since their ancestors were in a state of incomplete speciation (see Figure 6-9). Should the hybrid *Sabatias* be considered a separate species in their own right, or should they be ignored? In other words, is the presence of the hybrid *Sabatias* a valuable component of the biodiversity of the Hyannis Ponds, or is it an ephemeral accident that is of no consequence to the conservation of biodiversity? As of this time, these hybrids are the only report of any natural hybridization of any kind within the entire genus of 17 species. If the Hyannis Ponds population of hybrids is the only population in the world, then these are among the rarest plants in the world. Yet neither the Commonwealth of Massachusetts nor the local chapter of the Nature Conservancy has proposed that the hybrids be put on an endangered species list, or even that an individual hybrid is as valuable as an individual pure Plymouth gentian growing next to it, since the latter species is officially listed as a species of special concern in Massachusetts. That this story of the hybrid *Sabatia* exemplifies the problem of incommensurability can be shown by asking the following questions: To which species is the hybrid *Sabatia* commensurable? Are the two parent species, *S. kennedyana* and *S. campanulata*, somehow different as species from other nonhybridizing *Sabatias* because of the fact that they can produce hybrid offspring? Is this complex of *Sabatias* comparable as producers of hybrids to the complex of oaks, assuming the oaks are in the process of speciating and the *Sabatias* are not?

The world is teeming with examples of species that lead one to doubt that "a species is a species is a species." Consider *Poecilia formosa*, a species of guppy that consists entirely of females, whose eggs must be penetrated by the sperm of any one of three related bisexual species in order for development to occur (this condition is known as pseudogamy). No genetic fertilization takes place between the sperm of the bisexual species and the egg of *P. formosa*, and inheritance is purely maternal. However, on rare occurrences, true fertilization occurs, and the triploid offspring exhibit some paternal characteristics, including, very rarely, being male.[16] With what other species is *P. formosa* commensurable?

Ensatina eschscholtzi, a salamander species that is prototypical of the "ring species," raises the puzzling problem of species whose populations will interbreed with their neighboring populations as they stretch around a geographic barrier in the center but whose populations at either end of the ring will not interbreed with each other.[17] As conservation decision makers, do we want to acknowledge greater priority to intermediate populations of this species than to random populations of nonring species? For if we lose one or more of the *Ensatina* populations, the ring may be broken, and this exemplar of a rare species type may be lost, as the populations on each side of the lost population may split into two "normal" species.

Keep in mind the injunction we all received from our second grade teachers when learning the principles of addition, that "You can't add apples and oranges." The metaphor is particularly apt here, because when we add up species, though "species" superficially seems to be a sound unit for the purposes of addition, this is precisely the injunction we violate—we are adding apples and oranges. In other words, we treat all species as if they were fungible, as if, in quantifying biodiversity, we were proceeding by the following methodological rule: "Take every member of the class *species* and call them interchangeable." But as we brought up in the sections on species richness in Chapter 4, this is exactly the opposite of what we, as conservation decision makers, want to do. We specifically want to know if a particular species is rare or endangered, if it is taxonomically isolated, or if it plays an important role in the spiritual life of a human culture or in the functioning of an ecosystem. Not only that, but a closer look at the definition of species leads us to the conclusion that, biologically, there may be different types of species. At the very least, some species are clearly demarcated regardless of the methodology applied to diagnose them, while others, like the oaks, are problematic according to one or all methodologies.

CONCLUSION: AMBIGUITY IS NOT ALWAYS BAD

In their introduction to *Keywords in Evolutionary Biology*, the editors, Evelyn Fox Keller and Elisabeth A. Lloyd, begin by setting forth the "traditional model for scientific language" as one in which "terminological ambiguity, uncertainty, and double entendre are generally seen as evidence of scientific inadequacy—as impediments simultaneously to progress and to truth and, accordingly, as impurities requiring removal."[18] They continue by pointing to a more recent tradition, beginning with Kuhn in 1962, that argues that the goal of unambiguous scientific language not only is impossible but may not always be desirable for the advancement of science. Still, Keller and Lloyd conclude, "the very extent to which scientists . . . *aim* at a language of fixed and unambiguous meanings constitutes, in itself, one of the most distinctive features of their enterprise."[19] Thus, there is a tension, one that we wish to stress, between the need for precision and the impossibility of precision. The proper response to this tension is not, however, to give up our attempts for precision in language because of the profound difficulties of reaching that goal. The proper response is to be careful to articulate explicitly the range of meanings of a significant ambiguous term with each use, being especially careful when two or more uses of the same term are closely related but subtly different from each other. This is true for *diversity, biodiversity*, and *values*, and it is true for *species, ecosystems*, and *genes*. The proper response is also to recognize that we cannot treat ambiguous statements as if they were unambiguous, as if the statement "there are 16 species in one ecosystem and 12 species in a second ecosystem" is a precise scientific statement. Some of the species may be diagnosed differently from others; some of the species may exhibit boundary problems in space or time that others do not; the ecosystems may be of scales very different from each other. This does not mean that one should never say "there are 16 species in one ecosystem and 12 species in a second ecosystem"; it means that anyone who says it has the responsibility to acknowledge in what ways the inherent ambiguities of that statement affect the validity of his or her conclusions.

A retelling of the parable of the date farmers may be appropriate here. We borrow this parable from a paper by David L. Hull in which he applies the parable to the use of theoretical, nonoperational terms in science.[20] The palms are a family of flowering plants that include some 220 genera and 2500 species. There is a wide variety of palm fruits, including the betel nut, the coconut, and the date. The edible dates are harvested from a number of species, all within the single genus, *Phoenix*, although, like the oaks, the species boundaries are not well defined in this genus. A second important characteristic of date palms is that they are *dioecious*, which means that, like most higher animals, and unlike many other plants, individuals are exclusively either male or female (see Figure 6-10). The parable has it that early date farmers noticed that only half of their trees produced fruit. In order to

Figure 6-10.
Dioecy in palms. Pistillate (female) and staminate (male) flowers of the palm genus *Phoenix* (the date palms). Individual trees have either all pistillate or all staminate flowers. (After N. W. Uhl and Dransfield, J., *Genera Palmarum* [Lawrence, Kansas: Allen Press, 1987], 216)

increase the number of dates they could harvest from a given plot of land, they began to weed out the "sterile" (i.e., male) trees. This strategy proved successful, until the last male tree was cut down. All of a sudden, all the previously fertile trees stopped producing dates. Ambiguous terms are like the male trees. It is necessary to have them, and it is necessary to know which ones they are. Ambiguous terms are productive only when their numbers are kept low, when they are easy to recognize, and when the ambiguity they express is consonant with the ambiguity of nature.

References

1. R. K. Wayne and J. L. Gittleman, "The Problematic Red Wolf," *Scientific American* July (1995): 36–39.

2. Roger Lewin, *Human Evolution: An Illustrated Introduction* (Cambridge, Massachusetts: Blackwell Science, 1993).

3. F. A. Bazzaz and T. W. Sipe, "Physiological Ecology, Disturbance, and Ecosystem Recovery," in *Potentials and Limitation of Ecosystem Analysis*, ed. E.-D. Schulze and H. Zwölfer (Berlin: Springer-Verlag, 1987).

4. Robert M. May, "Levels of Organization in Ecology," in *Ecological Concepts*, ed. J. M. Cherrett (Oxford: Blackwell Scientific Publications, 1989), 346–347.

5. Gordon H. Orians, "Endangered at What Level?" *Ecological Applications* 3, no. 2 (1993): 206–218.

6. M. Lynne Corn, *Ecosystems, Biomes, and Watersheds: Definitions and Use* (Congressional Research Service, 93-655 ENR, 1993).

7. M. L. Shaffer, "Minimum Population Sizes for Species Conservation," *Bioscience* 31 (1981): 131–134.

8. L. C. Frosst, "The Study of *Ranunculus ophioglossoides* and Its Successful Conservation at the Badgeworth Nature Reserve, Gloucestershire," in *Rare Plant Conservation*, ed. Hugh Synge (New York: Wiley, 1981), 481–489.

9. K. J. Gaston, *Biodiversity: A Biology of Numbers and Difference* (Oxford: Blackwell Science, 1996). See particularly Chapters 2 and 7.

10. For a general discussion see Ereshefsky, ed., *The Units of Evolution* (1992). To be on the cutting edge of the debate, see any recent issue of *Systematic Biology* or *Biology and Philosophy*. For reviews pertinent to conservation biology, see Chapter 3 of G. K. Meffe and C. R. Carroll, *Principles of Conservation Biology* (Sunderland, Massachusetts: Sinauer Associates, 1994), and

Martha Rojas, "The Species Problem and Conservation: What Are We Trying to Protect?" *Conservation Biology* 6 (1992): 170–178.

11. Ernst Mayr, "Speciation Phenomena in Birds," *Am. Nat.* 74 (1940): 249–278.

12. See Mark Ridley, *Evolution* (Boston, Massachusetts: Blackwell Scientific Publications, 1993), 383–407.

13. A. J. Richards, *Plant Breeding Systems* (London: Allen & Unwin, 1986), 427–432; Michael J. D. White, *Modes of Speciation* (San Francisco: W. H. Freeman, 1978), 291–297.

14. Lucinda A. McDade, "Hybridization and Phylogenetics," in *Experimental and Molecular Approaches to Plant Biosystematics*, ed. P. Hoch and A. G. Stephenson (St. Louis: Missouri Botanical Garden, 1995).

15. Bruce Sorrie and Mario DiGregorio, personal communication, 1995.

16. Michael J. D. White, *Modes of Speciation* (San Francisco: W. H. Freeman, 1978), 296–297.

17. Ernst Mayr, *Animal Species and Evolution* (Cambridge, Massachusetts: Harvard University Press, 1963), 508–512.

18. E. F. Keller and E. A. Lloyd, *Keywords in Evolutionary Biology* (Cambridge, Massachusetts: Harvard University Press, 1992), 1.

19. E. F. Keller and E. A. Lloyd, *Keywords in Evolutionary Biology* (Cambridge, Massachusetts: Harvard University Press, 1992), 3.

20. David Hull, "The Operational Imperative: Sense and Nonsense in Operationism," *Syst. Zoo.* 17 (1968): 438–457.

7

Inventories

During our class field trip to Martha's Vineyard at the very start of the school year, we give our students an assignment: decide which of two patches of forest has more biodiversity. At Cedar Tree Neck, on Martha's Vineyard, we ask them to compare the forest on two sides of a path; on one side is a second-growth beech-maple forest, on the other is a small red pine plantation. Inventorying tasks, such as this assignment, are the starting point for any conservation action, for without information, even incomplete information, no intelligent conservation decisions can be made. We have discussed many potential and actual conservation decisions in this book, such as the taking of the Hyannis Ponds, whether or not to eradicate the impatiens from the Monteverde Cloud Forest Preserve, and how much effort to expend on protecting tuataras. Not one of these, nor any of the other decisions that we discuss throughout this book, could be taken without data about the real world—data that are gathered by inventories. Without accurate information about the current and historic ranges of the organisms in question, we would not know that the impatiens was alien to Costa Rica or that the tuatara is restricted to a few of New Zealand's smaller islands.

When we give the students this inventory assignment, we know full well just how subtle and difficult this task is—but they do not. As they settle down to their work, we settle down under a tree and wait for the questions to begin. "Should we count this plant that is half in and half out of the plot as part of the inventory? What about this seed? Something just burrowed into the soil, and I think it was a beetle; does that count?" They find, after a few hours, that they have made little headway toward completing their task, but they have developed a long list of important questions about what exactly they should be doing and how. Moreover, our students learn that the process of inventorying biodiversity is a value-laden exercise, that it involves choices and decisions about what will be the boundaries of their inventory in the geographic, taxonomic, and temporal landscapes that we discussed in Chapter 5.

PERSPECTIVES AND FILTERS IN PHOTOGRAPHY AND BIODIVERSITY INVENTORIES

Conservation decision makers draw the boundaries that shape their view of the world. We have found that a powerful analogy for these decisions is the set of choices that photographers make in creating each of their images. Where the photographer selects lenses, perspectives, and filters, the conservationist decides on the scope of an inventory. The world itself does not change in either case, merely the way it is perceived.

A photographer is an active, not passive, recorder of reality. Whether highly skilled professional or amateur snapshooter, the photographer makes numerous decisions about how to portray reality with every photograph. With even the simplest camera, the photographer chooses a place from which to shoot—which affects perspective and lighting—as well as selecting the framing of the subject and the moment at which to take the photograph. More sophisticated cameras offer the photographer additional choices of shutter speed and depth of field, focal length of lens, film, and the option of using filters. Perspective and filters are the two most powerful metaphors for thinking about biodiversity inventories from this list, and we will concentrate on them.

By changing position or lenses, the photographer has control over how subjects in the final photograph are viewed. By bringing the camera closer to the subject (or by using a telephoto lens), individual items within a larger scene can be emphasized. Conversely, by moving away from the subject or using a wide-angle lens, the photographer includes more of a scene in the photograph, presents a broader view, and de-emphasizes individual details (see Figure 7-1).

In the realm of biodiversity inventories, one can similarly emphasize certain features of the natural world. With such inventories, however, one tends to emphasize or de-emphasize entire classes of items, much as the photographic filters on a camera lens do. Such filters do not alter reality, but they alter the viewer's perception of reality and the film's recording of reality. By filtering out specific wavelengths of light, a filter removes certain aspects of visual reality; in doing so, the filter gives added prominence to those that remain.

So, too, a conservation decision maker may choose to ignore vast swaths of biodiversity in planning an inventory (or after an inventory has been created, as we discuss in Chapter 8). Only rarely does genetic diversity appear in an inventory, and frequently ecosystem diversity is left out of inventories as well. Considering diversity of species only, most taxonomic groups are filtered out of inventories, as invertebrates, microorganisms, and fungi only rarely show up in inventories (see Figure 7-2).

THE BASIC INVENTORY

The starting place for our students' work, and the starting place for many conservation decisions, is the basic field inventory. The definition of inventorying, according to the compendium *Global Biodiversity Assessment*, is the following:

> *Inventorying* is the surveying, sorting, cataloguing, quantifying and mapping of entities such as genes, individuals, populations, species, habitats, biotypes, ecosystems and landscapes or their components, and the synthesis of the resulting information for the analysis of processes.[1]

Figure 7-1.
Changing perspectives. By selecting different lenses, the photographer can either show a broader scene or emphasize specific details within a scene—just as a conservation decision maker can do. The photographs shown below were shot using lenses with the following focal lengths: (A) 35 mm, (B) 150 mm, and (C) 600 mm. (Photographs by Dan L. Perlman)

A

B

C

Figure 7-2.

"Filters" in inventories. Most aspects of biodiversity are "filtered out" of the inventory process and do not show up in the final inventory. In this diagram, only the information listed on the right passes through the filters. No information about genes and many groups of species is collected in the inventory, and little information about ecosystems appears.

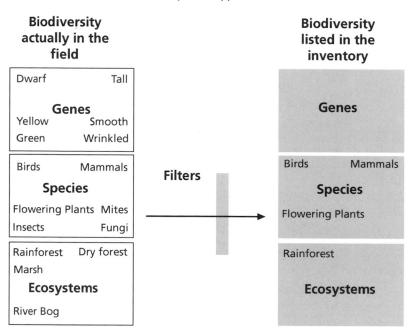

The underlying goal of creating an inventory is to answer the question, "What is here?" by recording the biological entities found at a site. As our students discover, each of the three simple words in the question, "What is here?" needs to be analyzed carefully before any of the more sophisticated words in the *Global Biodiversity Assessment* definition above can be put into practice. Furthermore, inventories are not objective; they are highly biased instruments, as we will see by examining each of the words in the question, "What is here?"

"Here"

"Here" clearly signifies place, an area in the geographic landscape. The key difficulty with the word is in deciding where to place the boundary between "here" and "not here." When we stand before our students and wave in the direction of the red pine plantation, we do not indicate the limits of the patch of forest to be considered. One of the first questions that the students come up with is "Where do we place the boundary between what is considered a part of the pine plantation survey and what is not?" The boundary could lie entirely within the tiny pine plantation, along what seem to be the edges of the plantation, or somewhere in the beech-maple forest beyond the plantation (see Figure 7-3). Clearly, each of these regions could be defined as

Figure 7-3.

The study site at Cedar Tree Neck Reserve. (A) The exact placement of the study plot on the plantation side of the path will yield very different sets of inventory data, as shown by the location of plots 1, 2, and 3 in the map. The location of plot 4 is less sensitive, as it is surrounded by extensive beech-maple forest. (B) The boundary between the red pine plantation and the beech-maple forest at Cedar Tree Neck, showing how plot 2 in the sketch map might appear. (Photograph by Dan L. Perlman)

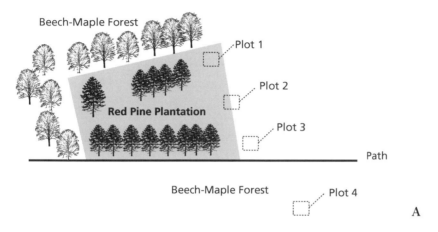

"here," and the choice of boundary would greatly affect the eventual assessment as to "what is here."

Consider again the photography metaphor introduced earlier. A photographer makes decisions similar to those of the biodiversity inventory planner. By selecting a certain lens, location, and angle from which to photograph, the photographer chooses to include a certain set of visual elements in the photograph and excludes all others—just as the conservation decision maker chooses to include a certain set of elements of biodiversity in the inventory and to exclude all others. And as with photographic choices, the conservationist's decisions about where to place the geographic boundary between what is in an inventory and what is not are neither right nor wrong. Instead, each decision must be calibrated against the goals, values, and interests of the conservation decision maker or photographer.

If the decision maker wants to know, "Which has more plant species, a pine plantation or a beech-maple forest?" then the pine plantation survey area should not include any beech-maple forest (just as the beech-maple site should not include any pine plantation; see plots 1 and 4 in Figure 7-3). On the other hand, if the goal were to determine which side of the path held a more representative sample of the region's biodiversity, the boundaries might be drawn differently. Without clearly articulated goals, the boundary-drawing process is essentially arbitrary. In fact, the boundaries that conservation decision makers work with usually are arbitrary. For example, the property lines of land that is under consideration for conservation are frequently due to historical accidents of fate and have little biological meaning. As such, the boundaries for inventories can be easily and clearly determined—they lie wherever the edge of the properties lie. But when attempting to survey a large, undeveloped region to decide where best to place a new nature reserve, the important issue of boundary selection is very much in the hands of the conservation decision maker.

"What"

"What" refers to the elements of biodiversity found at a site. But even in a pure red pine plantation with little undergrowth, answering the question of "what" is present requires a great deal of work and thought. Among our students' first questions as they begin their inventories at Cedar Tree Neck are "What should we be focusing on? Should we include the insects in our survey, since most of them are getting away and we can't identify them? Should we include the mosses, which we can't really tell apart?" Their questions do not include more difficult problems, such as what to do about inventorying genetic diversity. They know that with the current state of technology, it is impossible to assess the genetic diversity of a site; at most, a small part of the genetics could be learned about a few species at significant expense. So genetic diversity, although a part of the standard definition of biodiversity, is never really considered when answering "what is here?"

Just as it is not possible to assess all the genetic diversity at a site, nor is it possible to record all of the species present. Biologists still know so little about bacteria, fungi, and many groups of invertebrates (especially nematodes or roundworms, mites, and larval insects)—most of which are difficult to locate and identify—that it simply is not technically feasible to survey all of the species in any geographic area, even if that area is a red pine plantation less than a hectare in extent. James M. Tiedje, a microbial ecologist, has noted that studies "have indicated that there may be as many as 10,000 species of bacteria in a single gram of temperate forest soil . . . yet only about 3,000 species of bacteria have actually been named." Virginia Ferris, a nematologist, adds that in temperate forests "one can find 2 to 6 million nematodes per square meter, with as many as 200 species present—most of them unnamed and unknown"[2] (see Figure 7-4). Because an inventory cannot include all elements of biological diversity, not even all species, conservation decision makers must be explicit about which aspects of biodiversity should be included in an inventory and which should not. In other words, just as they drew geographic boundaries, they must also decide where to draw boundaries in the taxonomic landscape and actively decide what to include in a species-level survey.

Yet a further question comes up when surveying the biodiversity of a site; should the surveyors map (or otherwise note) the presence of biologically distinct communities and ecosystems? This step is different from merely listing the species found at a site; it calls for recognition of a different level of biodiversity. Not all regions have well-established classifications of communities and ecosystems, and the surveyor may either have to develop one from scratch or select from among several conflicting ones.[3] Clearly, such a process brings with it problems, such as knowing where to draw boundaries and what scale to use for describing different communities, issues we discussed in Chapters 5 and 6.

In practice, conservation decision makers select objects on which to focus, much as the photographer does (see Color Plates 10 and 11). Decision makers choose one or more target groups on which to focus, and ignore or filter out information about other groups. The focal groups are often delimited taxonomically (e.g., birds, bumblebees, or flowering plants), although they can be chosen on the basis of structure (e.g., trees with a diameter at breast height greater than 10 centimeters).

Finding high species richness in a certain taxonomic group at a site may not be indicative of species richness levels of other groups at that site. In a study of insects and birds living in forest fragments in Costa Rica, Daily and Ehrlich found that species richness in one group, such as butterflies, was generally not a good predictor of species richness in another group, such as moths. Specifically, they found no significant relationship between richness of butterflies and moths, between moths and birds, between beetles and birds, and between beetles and butterflies (see Table 7-1).[4] This point can generally be appreciated by studying tables and graphs of species richness in different groups, such as those found in *Global Biodiversity*.[5] Similarly, finding large numbers of endemic species from one group is not a good predictor of large numbers of endemics in other groups, as Prendergast and his colleagues found for several plant and animal taxa in Great Britain.[6] In our view, no particular focal group or taxon is "objectively" better than any other for assessing species richness across all groups at a site.

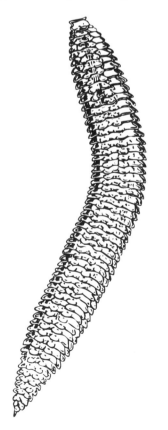

Figure 7-4.

A nematode, or roundworm. Temperate forest soils can hold 2 to 6 million of these creatures per square meter, yet few people ever become aware of these animals. (After R.C. Brusca and G.J. Brusca, *Invertebrates*. [Sutherland, MA: Sinauer Associates, 1990:351])

Table 7-1. Correlations of species richness of different taxa in forest fragments.

	Moths	Beetles	Birds
Butterflies	No	No	Yes
Moths		Yes	No
Beetles			No

"No" means there was no significant correlation in species richness between the two groups. "Yes" means that a significant correlation in species richness was found between the two groups.
SOURCE: Adapted from G. C. Daily and P. R. Ehrlich, "Nocturnality and Species Survival," *Proceedings of the National Academy of Sciences*, 93 (1996):11709–11712.

The selection of which groups to include in an inventory depends greatly on the values and goals, as well as the budget and expertise, of the conservation decision makers asking for the survey. Foresters rarely call for inventories of salamanders, and few organizations have either the interest or the funds for extensive inventories of insects. Target elements are usually selected for a combination of ecological and economic importance and ease of surveying. In general, woody plants are highlighted in inventories. Woody plants have the advantage of being relatively permanent members of ecological communities. In addition, they are useful indicators of underlying soil and climatic conditions, which constrain the types of plants and animals that live in a region.

"Is"

In many ways, the issue of what "is" found at a site is the most difficult to resolve. The word "is" implies the present, but the question that should be asked is "What is, was, or will be here?" However, our students find some of the problems with this term when they ask questions such as, "Should we count that ant that just walked across the corner of our study plot as being present or not?" and "I think I saw a bird fly overhead; does that count?" In practice, conservation decision makers do not want the results of just a single inventory to determine "what is here." Several types of variation over time can affect the results of an inventory, and only multiple inventories allow the decision maker to understand such variation. Conservationists must also decide on how to interpret variation over time and what kinds of boundaries to draw in the temporal landscape.

Annual and Seasonal Variation.   If our team of students surveyed the forests of Cedar Tree Neck during February, they could easily find and identify the tree species present, but many species of small warm-weather herbs would not be visible, a large proportion of the bird fauna would have headed south for the winter, and most insect species would be hidden in overwintering sites. In fact, a February inventory of Cedar Tree Neck would miss most of the site's elements of biodiversity other than large woody plants—part-time residents would be gone, and most year-round residents would be hidden from view.

Because a single survey misses so much, the inventory team might want to return to Cedar Tree Neck on a regular basis, perhaps every two or three weeks throughout a year. With this type of scrutiny, virtually all of the herbs and birds of the forest will be seen sooner or later (although the team would need to perform nighttime inventories to record the owls and nocturnal insects of the site). Many insects would be visible at one time or another, although a good proportion of cryptic and subterranean species might escape the team's notice. But here a different problem comes up: What are the minimum number and types of observations necessary for a species to be considered as present at the site?

"Just Passing Through" Versus "Present." The standard for "presence" will likely vary among groups of organisms and according to the goals of the inventory. A single sighting of a tree is qualitatively different from a single sighting of a bird—the tree can be found again by surveying the patch of forest where it was first seen, whereas the bird may have been a transient visitor. Is a single observation of a bird at an intensively studied site sufficient to consider that bird a resident of the site? Or are multiple observations or evidence of nesting behavior a more useful standard?

As happens so often in assessments of biodiversity, apparently straight-forward facts frequently require interpretation. For example, Dick Johnson, executive director of the organization that owns the reserve, the Sheriff's Meadow Foundation, told us, "The worm-eating warbler has been seen at Cedar Tree Neck once in the past five years." If a conservation decision maker wants to decide whether the worm-eating warbler should be considered "present" in an inventory of Cedar Tree Neck, this data point must be actively interpreted relative to the goals of the conservation decision maker. Different standards apply if the conservation decision makers are looking for breeding populations of birds or merely interested in seeing what migrant species pass through the site. Moreover, the quality of the data has to be considered: Who was the observer, and how was the observation made? How sure are we that the observed bird was even a worm-eating warbler at all?

Natural Variation Between Years. Inventory results can vary between years as well as within years—often quite drastically. The first year that the authors visited Cedar Tree Neck to evaluate it as a possible field site for our classwork, we were overwhelmed by the quantity of poison ivy (*Toxicodendron radicans*) growing in the forest. Neither of us had ever seen anything like it; the species dominated the understory and herb layers of the forest. In fact, there was so much of the noxious plant about that we considered not bringing our students to the reserve. Because, in all other respects, it was a perfect site for our needs, with its forest, rich ponds, and seashore, we decided that we should bring our students there, and have done so for several years. The second year we visited the site we were just as surprised as the first—the poison ivy that had been everywhere had suffered a population crash. We estimated that there was only 2% to 5% as much poison ivy present the second year as there had been the first year. Again, neither of us had ever seen anything like such a crash. We discussed the phenomenon with director Dick Johnson, who confirmed that the population crash was natural and who was just as surprised about it as we were.

Had we performed an extensive series of biodiversity inventories during the first year and written up the results, we would have spent a significant portion of our report describing the "poison ivy-beech-maple community," whereas a report based on a similar series of inventories during the second year would have had little to say about poison ivy. From one year to the next, the single most striking aspect of the vegetation had changed from being a dominant and spectacular feature of the landscape to something quite ordinary—and this change involved a native species. The situation can be even more extraordinary when an invasive exotic species appears on the scene. Weedy species such as kudzu and fire ants in the southern parts of the United States and purple loosestrife in wetlands throughout the eastern parts of the country can fundamentally alter the biological makeup of a site in just a few years. A careful multiyear survey can become completely obsolete in just a few years if an exotic species sweeps through the region after the last site visit.

Succession.   The issues discussed so far concerning "is" have all been relatively short-term, spanning periods of months or a few years. Perhaps the most fundamental issue, however, concerns succession—the changes that occur at a site as it matures over time, typically after a disturbance of some kind. A Vermont forest that today contains mostly white birches, aspens, and red maple trees will probably look very different in an inventory taken 100 years from now, after the later-successional species, such as sugar maple, basswood, and yellow birch, have begun to grow on the site. Conservation decision makers must take note of potential succession when planning nature reserves and management plans, as we discussed in Chapter 6.

Thus, the apparently simple process of creating an inventory for a site in order to answer the question, "What is here?" turns out to be a complex procedure that requires planning, decision making, and interpretation of the results. In the next section we examine two of the most ambitious inventories currently being performed: the nationwide biodiversity survey of Costa Rica's National Biodiversity Institute and the multiyear, multi-million-dollar biological inventory in the Costa Rican province of Guanacaste.

THE GRANDEST OF ALL INVENTORIES: COSTA RICA'S NATIONAL BIODIVERSITY INSTITUTE AND THE ALL TAXA BIODIVERSITY INVENTORY

A massive amount of work is needed to fully catalog the biodiversity of a region. If one were to employ the standard definition of biodiversity, as in the OTA definition quoted in Chapters 1 and 2, and attempt to catalog every gene, species, and ecosystem in a region, what kind of effort would be needed? History is no guide, because such an inventory has never been compiled. However, two landmark efforts in Costa Rica stand out as the most ambitious ever attempted: the All Taxa Biodiversity Inventory and the national inventory sponsored by Costa Rica's National Biodiversity Institute (INBio).

The INBio effort, begun in 1989 under the direction of Rodrigo Gamez, is a countrywide attempt to catalog the biodiversity of Costa Rica. INBio employs systematists to survey most major groups of organisms, paying special attention to populations in nature reserves throughout the country.[7] As part of this program, INBio has trained several dozen local Costa Ricans as para-taxonomists, individuals who participate in a training program that teaches them how to distinguish, in the field, the diversity of a wide range of taxonomic groups, especially plants and invertebrates, without having to go through the 8 to 10 years of training of a Ph.D. systematist. Daniel Janzen, a biologist whose work has included ecology, systematics, and conservation throughout the New World tropics, especially in the Guanacaste Province in northwest Costa Rica, has worked with Gamez throughout the inception of INBio.

In conjunction with INBio Janzen began an All Taxa Biodiversity Inventory (ATBI), a more intensive inventory of the Guanacaste Conservation Area. The ultimate goal of this ATBI was to collect, describe, and gather information on every species of every kind—animals, plants, bacteria—that is found in that area of 1200 square kilometers (see Figure 7-5). As INBio states on its World Wide Web site, "An ATBI aims to find out what species are there, where they are, what is their basic natural history, and how to make them available and useful."[8] Janzen has estimated that, after a two-year

Figure 7-5.
Map of Costa Rica. Note the Guanacaste Conservation Area, site of the five-year, $100-million All Taxa Biodiversity Inventory.

planning period, the species and ecosystem inventory process would take five years at a cost of about $100 million.[9]

While this book was in press, the ATBI was terminated after four years of planning.[10] The people most involved in planning and running the ATBI wrote a letter on the World Wide Web "To all [who] have been involved in the All Taxa Biodiversity Inventory (ATBI) of Costa Rica's Area de Conservación Guanacaste (ACG)" that included the following:

> [W]e have come to the conclusion that the necessary political economic and institutional conditions to actually conduct an ATBI of the ACG are not present, either nationally or internationally. It is not the right combination of the time, place and circumstances in the history of tropical biodiversity development to carry out this initiative. An ATBI may well appear somewhere in the future but it would be non-productive to continue to try to force it to appear here now.[11]

Despite its termination, the ATBI as planned is worth considering for what it says about the role of values in creating inventories. We expect that this concept will surface again, and so we write about it in the future tense, assuming that a future ATBI would be much like the one that was planned for Guanacaste. INBio itself continues its mission of inventorying the biodiversity of Costa Rica and will continue to do so.

Like all inventories, the INBio effort and the ATBI are products of their developers' values and interests, as well as the constraints within which they operate. To a greater extent than usually found, both Gamez and Janzen have been explicit about their values.[12] Janzen, for example, has written: "An inventory is not an end but rather a means to the end of efficient use of a repository of diverse objects too complex to be censused at a glance."[13] His ends of efficient use are varied, but clearly included as among the most important are the extractive and nonextractive uses that can benefit the lives of those people living in closest contact with natural areas in the tropics.

Traditional university systematics research and the All Taxa Biodiversity Inventory are similar activities in many ways—both attempt to collect, identify, and name species from the wild. But, as Janzen points out, the values that are driving these two research programs can be quite different. University taxonomists usually work to produce comprehensive monographs of a group of closely related species, and oftentimes those species can be found in areas of the globe quite distant from each other. University taxonomists also tend to produce keys (that is, identification aids) to the species that are based on technical characteristics with which only trained biologists would be familiar. The ATBI researchers, on the other hand, will produce local lists of species accompanied by pictorial keys that use easy-to-understand characteristics. Though evolutionary relationships between species will not be ignored, the ATBI will pay far more attention to gathering information about the natural history and potential economic uses of the species, information that will be made available to a variety of people, including local *campesinos*, agriculturalists, educators, scientists, foresters, conservationists, and industries such as eco-tourism, pharmaceuticals, and biotechnology.[14]

A different aspect of values also comes into play in the choice of which groups of organisms the ATBI will put its resources into cataloging. Although the stated goal of the ATBI is to catalog "what species are there" within the study area, the protocol for the project creates a hierarchy of groups, in light of the acknowledged inability to know all of biodiversity. The task of appending useful scientific names and basic ecological information to the species in the Guanacaste Conservation Area will be accomplished by "taxonomic working groups" (TWIGs), collections of professional academic systematists, graduate students, and parataxonomists. For the purposes of an ATBI, the protocol divides the TWIGs into four groups based on the possibility of complete identification to species level within the time and budgetary constraints of the ATBI.[15]

The Group 1 taxa are manageable taxa that can be almost totally identified to the species level within the five years of the ATBI, including vertebrates, Odonata (dragonflies and damselflies), lichens, ferns, mollusks, centipedes, tapeworms, and vascular plants. These species constitute less than 10% of the total ATBI. Group 2 taxa, including spiders, Homoptera (aphids and their relatives), primitive plants, animal and domesticated plant viruses, worms, and diatoms, are those taxa that will have most of their organisms identified to the species level during the five-year ATBI. More species may come to light after the official close of the ATBI, but not in significant numbers.

Group 3 taxa are species-rich taxa from which a significant fraction of species will remain unknown and uninventoried at the end of the ATBI. These species are difficult to collect, relatively unknown, and lack higher taxonomic features that make them easily distinguishable. Species from taxa such as bacteria, fungi, micro-Hymenoptera (tiny wasps), mites, prokaryotic organisms, nematodes, and protists will fall into this category. Within these taxa, many subsets will be identified and collected thoroughly enough to join the group 1 or group 2 species, but others will end up classified as group 4. Taxa designated group 4 will not receive taxonomic efforts during the course of the ATBI. Often they are so-called "orphan taxa," groups without available experts. Groups such as wild plant viruses, rotifers, annelid worms, and collembola will probably remain unstudied. Specimens will, however, be collected and saved for later research.

Accordingly, even in an *all*-taxa inventory some taxa will be completely left out. This is mandated by limited resources—limited money, limited time, and limited expertise. But the limited expertise—that is, the fact that few individuals have ever studied some groups—is a reflection of societal values larger than those of the originators of the ATBI. Few people have studied the systematics of nematodes, both because this group appeals to far fewer people than do groups such as birds, butterflies, and flowering plants, and because society has found fewer reasons to reward people to engage in such studies.

So it is that when our students begin to compare the "biodiversity" of two sites early in the fall, they never stop to consider the diversity of the nematode fauna—many of them enter our class never having heard the word "nematode." When we direct our students to inventory their sites, they come up with many questions that they must answer before they can complete, or even begin, this task. Most of their questions, which could be asked about any inventory, cannot be answered in the abstract. Instead, the questions must be considered in relation to the decision maker's overall goals for the inventory. At a larger scale, the focus of any inventory must be shaped by the conservation priorities held by the conservation organization creating the inventory. That the ATBI could not actually be started, despite the fact that approximately $25 million had been secured or was strongly expected for project funding, is an indication that there may have been differences among the values, goals, and priorities of the people planning and sponsoring the inventory.[16] It is to the subject of goals and priorities that we turn in the next chapter.

References

1. United Nations Evnironment Programme, *Global Biodiversity Assessment* (Cambridge: Cambridge University Press, 1995), 459.
2. Carol K. Yoon, "Counting Creatures Great and Small," *Science* 260 (1993): 620–622.
3. Robert G. Bailey, *Ecosystem Geography* (New York: Springer-Verlag, 1996).
4. G. C. Daily and P. R. Ehrlich, "Nocturnality and Species Survival," *Proceedings of the National Academy of Sciences*, 93 (1996): 11709–11712.
5. World Conservation Monitoring Centre, *Global Biodiversity: Status of the Earth's Living Resources* (London: Chapman & Hall, 1992); International Council for Bird Preservation, *Putting Biodiversity on the Map: Priority Areas for Global Conservation* (Cambridge: International Council for Bird Preservation, 1992).
6. J. R. Prendergast, R. M. Quinn, J. H. Lawton, B. C. Evershame, and D. W. Gibbons, "Rare Species, the Co-incidence of Diversity Hotspots and Conservation Strategies," *Nature* 365 (1993): 335–337.
7. United Nations Evnironment Programme, *Global Biodiversity Assessment* (Cambridge: Cambridge University Press, 1995), 520.
8. http://www.inbio.ac.cr/ATBI/ATBI.html on July 29, 1996.
9. http://www.inbio.ac.cr/ATBI/ATBI.html on July 29, 1996.
10. http://www.inbio.ac.cr/ATBI/ATBILetter.html on 13 June 1997. Letter dated 8 November 1996.
11. http://www.inbio.ac.cr/ATBI/ATBILetter.html on 13 June 1997. Letter dated 8 November 1996.
12. Rodrigo Gamez, personal communications, several dates from 1990–1994.
13. D. J. Janzen, "A South-North Perspective on Science in the Management, Use, and Economic Development of Biodiversity," in *Conservation of Biodiversity for Sustainable Development*, ed. O. T. Sandlund et al. (Oslo: Scandinavian University Press, 1992), 38.
14. D. J. Janzen, "A South-North Perspective on Science in the Management, Use, and Economic Development of Biodiversity," in *Conservation of Biodiversity for Sustainable Development*, ed. O. T. Sandlund et al. (Oslo: Scandinavian University Press, 1992), 38–44; and http://www.inbio.ac.cr/ATBI/ATBI.html on July 29, 1996.
15. All Taxa Biological Inventory Protocol, Section 8.3.1.1 (1993).
16. http://www.inbio.ac.cr/ATBI/ATBINews2.html on 13 June 1997.

Articulating Goals and Setting Priorities

The Biodiversity Support Program, a conservation consortium sponsored by the U.S. Agency for International Development, held a workshop for biodiversity experts in 1994 to set geographic conservation priorities throughout Latin America and the Caribbean.* The format of this priority-setting exercise was in many ways a model for how we envision such discussions proceeding. During the first three days of the workshop, biodiversity experts divided all of Latin America and the Caribbean into seven major habitat types, which were further subdivided into a total of 35 regional habitat units (RHUs) (see Table 8-1; see also Figure 6-7 for map). Groups of participants with expertise in various taxonomic groups (plants, insects, birds, reptiles and amphibians, mammals, and freshwater fish) evaluated the entire region, one RHU at a time, and ranked each unit according to their perception of the area's biological worth for the taxonomic group in question. Each group considered criteria such as "species richness, phyletic diversity, number of endemic species, beta diversity and presence of rare/endangered species."[1] These different evaluations were then aggregated into a summary "biological value" score for each RHU.

At the same time that the expert groups were assessing biological value, participants assessed the "conservation status" of each RHU.† Based on their experiences in different regions, participants generated data on five features that were employed in assessing "landscape integrity and conservation status." These features were "1) the presence/absence of large blocks of original habitat; 2) the percent of remaining original habitat; 3) the rate of conversion; 4) degree of degradation and fragmentation; and 5) degree of protection."[2]

With the data collected, each RHU was assigned to one of five categories of conservation status, listed in decreasing degree of threat: critical, endangered, vulnerable, relatively stable, or relatively intact (see Figure 8-1).

On the final day of the workshop, four groups of participants were each "asked to devise a methodology to . . . arrive at a list of priority areas for biodiversity conservation."[3] Each group took as its starting point the biologi-

---

*The Biodiversity Support Program includes Conservation International, The Nature Conservancy, the Wildlife Conservation Society, the World Resources Institute, and the World Wildlife Fund.
†The participants began by assessing 148 "ecoregions," then assembled the data into the larger regional habitat units (of which there were 35) to ensure compatibility with the biological assessment.

Table 8-1.   List of the major habitat types and regional habitat units delineated in the Biodiversity Support Program study. See Figure 6-7 for a map of these units.

Major Habitat Type (MHT)	Regional Habitat Unit (RHU)
1. Tropical Moist Lowland Forests	1. Atlantic
	2. Upper Amazon
	3. NE Amazon
	4. SE Amazon
	5. Chocó-Darién
	6. Central American Lowland
2. Tropical Moist Montane Forests	1. Tropical Andes
	2. Central American Montane
	3. Caribbean Moist
	4. Venezuelan Coastal
	5. Guayana Montane
3. Tropical Dry Forests	1. Northern South American Dry
	2. Western Andes
	3. Chaco
	4. Central American Dry
	5. Mexican Dry
	6. Cerrado-Pantanal
4. Xeric Formations	1. Mexican Xerics
	2. Caribbean Xerics
	3. Caatinga
	4. Peru-Chile Deserts
	5. Chilean Winter Rainfall
	6. Argentine Monte
5. Herbaceous Lowland Grasslands	1. Central American Pine Savanna
	2. Llanos-Gran Sabana
	3. Pampas
	4. Patagonian Steppe
	5. Amazonian Savannas
6. Herbaceous Montane Grasslands	1. Paramo
	2. Puna
	3. Southern Andean Alpine
	4. Pantepui
7. Temperate Forests	1. Southern Temperate Forest
	2. Brazilian *Araucaria*
	3. Mexican Pine-oak

SOURCE: Biodiversity Support Program, Conservation International, The Nature Conservancy, Wildlife Conservation Society, World Resources Institute, and World Wildlife Fund, *A Regional Analysis of Geographic Priorities for Biodiversity Conservation in Latin America and the Caribbean* (Washington, D.C.: Biodiversity Support Program, 1995), 4, [Table 2].

cal value and conservation status data assembled during the first three days of the workshop. Each began with exactly the same data sets—yet the groups came up with different conservation priorities.

All four groups gave highest priority to those areas designated as containing the most "biological value" and lowest priority to areas with the least "biological value"; this result is not at all surprising. Three of the groups treated the conservation status data in fairly standard fashion, setting areas with critical conservation status as the top priority, followed by endangered,

Figure 8-1.

Levels of threat. During the Biodiversity Support Program workshop on conservation in Latin America and the Caribbean, participants developed a scheme for ranking ecosystems according to how threatened they are. The categories created are shown in this figure.

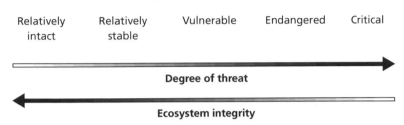

Figure 8-2.

Outcome of decision rules. Different values can produce different results. (A) Degree of threat was the only consideration used by three of the four groups at the Biodiversity Support Program workshop. (B) Both degree of threat and intactness of ecosystem were important considerations in the decision-making process of one of the groups.

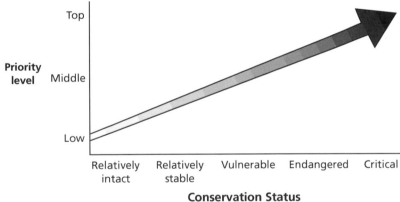

A

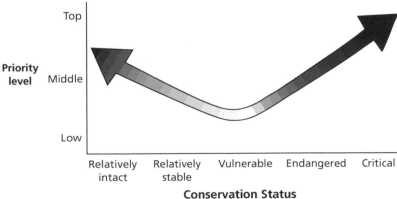

B

vulnerable, relatively stable, and relatively intact areas, in that order. This system of setting priorities is straightforward; where threat exists, a response is called for, and the size of response should be commensurate with the size of threat (see Figure 8-2). The fourth group took a different tack and considered relatively intact areas as well as critical ones to be of high priority. In a sense, they gave priority to habitat units in the best and the worst condition. They wrote, "relatively intact RHUs also deserve immediate conservation investment because it is within these now rare landscapes that ecosystem processes and species have the best chance for long-term persistence."[4] Different groups came up with different priority-setting strategies: What were the values underlying the thinking and priority-setting plans of the different groups?

Each of the groups employed two specific values in setting priorities, while the fourth group employed an additional value. The two universal values might be called the following, assuming that all else is equal among the areas being compared:*

- *Biological value:* Areas with greater biological value receive higher priority for conservation action than areas with lower biological value.

- *Degree of threat:* Areas at greater risk receive higher priority for conservation action than areas at lower risk (according to the conservation status scale).

Employing only these two values was relatively simple and involved no great conflicts. In contrast, the additional value used by the fourth group increased the complexity of the priority setting process. This value was:

- *Integrity:* Areas that are more intact receive higher priority for conservation action than areas that are less intact (according to the conservation status scale).

The degree-of-threat and integrity values are in direct conflict, as the former accords high value to the "at risk" end of the conservation status scale and the latter accords high value to the "intact" end of the same scale. Was this a problem for the group advocating both of these values? No—the group found a great deal of worth in *both* especially threatened and especially intact areas and less worth in areas that were only moderately intact (or alternatively, that were moderately threatened). This group articulated a value that emphasizes the worth of intact ecosystems and led to such areas being given high priority, something that none of the other groups advocated. In a world of scarce conservation resources, articulating and acting upon such a value might make a huge difference in the kind of biodiversity legacy that we leave behind, for unless large, intact areas are specifically given attention, they may soon become fragmented and degraded, as have so many other areas.

In our view, the two most striking aspects of this workshop were 1) that the working groups articulated their priority-setting goals and values so clearly, and 2) that the groups differed significantly from each other in their recommendations. By stating their goals and values clearly, the workshop participants have made a major stride forward in discussions of priority

*The wording of the values is our interpretation of the decision rules that the different workshop groups used.

setting; others involved in such activities can consider the goals stated by the Biodiversity Support Program workshop participants. That the fourth group's proposal differed so much from the others is an important reminder that multiple voices enhance the discussion of priorities.

In this chapter we discuss the ways in which the goals and values held by organizations shape and constrain the entire process by which they set conservation priorities. We examine the interactions between goals and values on the one hand and the priority-setting process on the other. For the purposes of this chapter we consider the priority-setting process as consisting of three steps:

1. the creation of biodiversity inventories (treated in Chapter 7);

2. the selection of specific elements of biodiversity to be targeted for protection based on inventory data (such selection procedures are discussed later in this chapter);

3. the development of action plans for protecting the selected elements of biodiversity (a topic, dealt with in most conservation texts, that is beyond the scope of this book).

We examine how values and goals interact with each of these steps, paying special attention to several different selection procedures that have been proposed.*

CENTRAL GOALS: DETERMINING WHAT TO PROTECT

Each priority-setting process is shaped by a large number of distinct values—but we have found that most methods embody one or two overriding values that we call the *central goals*. Every priority-setting process discussed in this chapter aims to protect biodiversity, yet because they can be based on such disparate central goals and values, they may lead to widely differing proposals for action—as we saw in the Biodiversity Support Program example. Central goals affect each aspect of the priority-setting process, including shaping the conservation organization's practical definition of biodiversity and the choice of what elements of biodiversity on which to focus.

We have found that most priority-setting processes embody one of four distinct central goals, although some contain more than one of these. The goals are:

1. protecting *all* of biodiversity;

2. protecting the "*most*" biodiversity;

3. protecting the *most diverse subset* of biodiversity;

4. protecting the *most valuable* biodiversity.

---

*Much of our early thinking for this chapter was stimulated by reading Nels C. Johnson, (1995), *Biodiversity in the Balance: Approaches to Setting Geographic Conservation Priorities*. Washington, D.C.: Biodiversity Support Program, a superb analysis of geographic priority setting.

In large measure, these four central goals represent answers to the question, "If I want to protect biodiversity, what should I do?" and the answers differ significantly from each other.

In considering the terms used in these four categories, a number of critical issues arise, and the conservation organization setting priorities needs to be explicit about what each term means. First, what does it mean to "protect" biodiversity, and what is the definition of "biodiversity" that the organization employs in practice? Each conservation organization probably has a different conception of what these two terms mean, affecting the types of inventories that it makes and the action plans it considers. Elements of biodiversity, such as domestic crop varieties, that do not even appear in an organization's practical definition of biodiversity will certainly not be considered for protection, and any protection that they do receive will be incidental. Second, how does one measure relative quantities of biodiversity, so that one can distinguish greater and lesser quantities of biodiversity in searching for the "most" biodiversity? Third, the conservation organization's concept of which aspects of biodiversity are "valuable" and "diverse" need to be made explicit. Finally, all of the concepts used by the organization must be compared with those employed by the developers of the specific selection procedures, since a lack of fit can cause problems. We next consider each of these four central goals and the issues surrounding them in detail.

Protecting *All* of Biodiversity

A number of priority-setting methods take this as their central tenet, including many of the oldest priority-setting methods. Noah, in the Flood story of Genesis, was expected to protect at least one mating pair of every animal species. The U.S. Endangered Species Act as it was originally passed in 1973 has a similar standard, protecting *all* organisms with the exception of insect pests. The Convention on International Trade in Endangered Species (CITES) includes in its highest category of protection "all species threatened with extinction which are or may be affected by trade."[5] (The limitation to species affected by trade derives from the fact that this is one of the most tractable areas in which to create international law.)

This central goal has a long history in the religious/moral realm and is based on the tenet that humans have a responsibility to protect all of life. Furthermore, this goal has a certain heroic feel to it, as one commits to protecting even the most obscure of species. In the real world of conservation action, however, this goal carries certain costs with it.

An important corollary of the injunction to "protect all" is the emphasis on attempting to "protect the most threatened" elements of biodiversity—for if we do not protect these elements, we are likely to lose them forever, thereby failing in the attempt to protect all. In practice, this is largely how the Endangered Species Act functions. By making degree of threat the paramount criterion in setting priorities, however, conservation organizations lose most of their flexibility to select criteria for action. Instead, their priority-setting task devolves into a never-ending assessment of threats to which they are forced to respond, and employing threat as the primary criterion for priority setting becomes a rigid constraint. In a world of limited resources for biodiversity protection, conservation organizations must ask whether they want to expend all of their resources dealing with emergencies as they arise, or whether they want to put some of their efforts into preventative measures. Threat is certainly an important factor, and we do not advocate ignoring it. However, all of the world's current conservation resources, both money and human expertise, are insufficient to deal with the number of species that are gravely threatened today (not to mention threats to genes and ecosystems and elements of biodiversity on other levels). Even the United States, the world's richest country, has hundreds of species waiting for official listing or development of recovery plans under the Endangered Species Act—and that does not include the vast numbers of obscure invertebrates that have never been adequately censused.[6] Once conservation organizations recognize that they cannot in practice "protect all," they must consider selecting a different central goal.

Protecting the "*Most*" Biodiversity

Attempting to protect the "most" biodiversity possible is a potentially reasonable goal, especially if conservation organizations have little knowledge of the particulars of, and context surrounding, assemblages of biodiversity elements. Proponents of this goal necessarily treat biodiversity as an entity that can be measured, that can be placed on a scale and weighed in some fashion. They assume that there is an unambiguous answer to the question, "Which of these assemblages contains more biodiversity?" and that humans can determine what that answer is.

A fundamental problem with this goal is whether one can "measure" biodiversity precisely and consistently in any meaningful way. Bryan Norton, philosopher of science, states that "most advocates of biodiversity protection assume that we have—or will have—a precise scientific measure of biological diversity and that the measure, once articulated, will tell us which elements of nature to 'target' in our conservation efforts," after which he discusses various arguments about why simple measures of diversity cannot be constructed.[7] We agree with Norton that biodiversity is such a complex entity that it cannot be measured and expressed as a single quantitative variable in the way that, for example, length, area, and mass can. Despite the technical difficulties of measuring the height of a tree, the mass of the Earth, or the distance of a galaxy, each of these measurements is taken in order to determine a single, well-understood quantity as accurately as possible; there is little argument over the definition or nature of the quantity being measured in each of these cases. Biodiversity is a complex concept, much like human health or human society, and "measurement" of such entities is an unsatisfactory enterprise. These entities possess large numbers of features to be measured, many of which are difficult to define conceptually and in practice, such as an individual's intelligence, a society's health, and the ecological niche of a species.

Individuals and organizations that subscribe to this goal rely heavily on species richness and allied measures of ecological diversity to determine priorities.[8] Species richness, however, is arguably the most sterile measure of biological diversity available, since it is based on inventories that have had *all* of their detailed information stripped away, leaving only a single, dimensionless number behind. Nonetheless, the goal of protecting the "most" biodiversity shows up frequently in selection procedures, especially those that acknowledge that they are merely first passes at determining priorities.

And when used in conjunction with other criteria such as integrity, endemism, rarity, and taxonomic isolation, species richness may prove useful in outlining rough-and-ready priorities.

One method that relies heavily on species richness as a priority-setting criterion is the "megadiversity country" approach (this method also includes degree of threat, endemism, and presence of charismatic mammals as criteria). Often when the biodiversity of different countries is being compared, species richness is the single most prominently displayed piece of information, as in the data tables of *Global Biodiversity* and *World Resources 1996–97.*[9] The authors of *Global Biodiversity* justify this practice by stating: "Although straightforward measures of species richness may convey relatively little ecologically important information, in practice because they are the most easily derived, they are perhaps the most useful index for comparisons of biological diversity on a large scale."[10]

Several Australian conservation biologists have developed methods for determining the most efficient patterns for protecting representative samples of the biodiversity found in a region.[11] These methods emphasize the criterion of representativeness (i.e., attempting to protect a sample of a region's biodiversity that best represents the entirety of the region's biodiversity) and the principle of complementarity (that each succeeding conservation effort should try to protect as much of the as-yet-unprotected biodiversity as possible, while passing over already-protected elements). This work, which contains sophisticated computer models, is founded on the central goal of protecting the most biodiversity, as can be seen in the following statement: "Reserves are most likely to fulfill their critical role in conserving biodiversity if . . . they contain as many elements of biodiversity as possible."[12]

Despite the role that species richness still plays, we recognize that most conservation decision makers are moving away from using such measures to identify areas for biodiversity protection, a change that we support.

Protecting the *Most Diverse Subset* of Biodiversity

A number of selection procedures have been developed that attempt to "maximize diversity" among the elements of protected assemblages. The developers of these procedures argue that if we cannot protect all of biodiversity, then we should protect as diverse a subset as possible. These procedures suggest methodologies for deciding which subset to protect when one only has sufficient resources to protect a small proportion of the biodiversity needing protection.

A group of systematists at the British Museum, who have developed several methods for identifying diverse subsets of taxa, note that the *Shorter Oxford English Dictionary* defines *diversity* as "difference, unlikeness." These authors, P. H. Williams, R. I. Vane-Wright, and C. J. Humphries, go on to write that "We have adopted a definition of biodiversity that includes not only the number of taxa in a sample, but also some measure of the *degree of difference* (in the sense of dissimilarity) among them" (emphasis in original).[13] We believe that this definition, especially if it is broadened to include elements other than taxa (e.g., genes and ecosystems), is an important step forward in thinking about biodiversity. All too often, biodiversity is defined in practice as a mere count of elements. In such cases, too little attention gets paid to the underlying differences among elements that are the essence of diversity.

Figure 8-3.
The blue whale. The largest creature ever to live on Earth has been known to hybridize with its closest relative, the fin whale. Is the blue whale less valuable because it has another species so closely related to it?

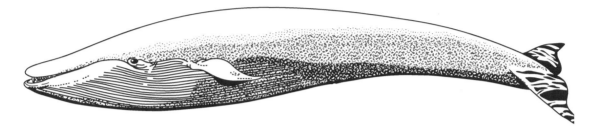

Having made the crucial leap of including the concept of difference and diversity in their definition of biodiversity, however, Williams et al. immediately settle upon a single measure of the diversity in biodiversity:

> Clearly many kinds of differences could be recognized between species. We have attempted to measure the overall differences between species in terms of their genealogical relationships. This approach is based on the assumption that, in general, patterns of difference among species are most likely to be congruent with the pattern of their genealogical relationships through genetic inheritance.[14]

Although many "patterns of difference among species" are "congruent" with their phylogenetic relationships, and a great deal of information about diversity among species is captured by phylogenetic trees, methods based on such trees also miss a great deal of the "diversity" that is found in assemblages of biodiversity. For one thing, intraspecific genetic diversity and ecosystem diversity simply do not appear in these schemes; only species and other taxa are recognized. For another, even at the species level a great deal of information is missed, as when one of several closely related species exhibits a single remarkable trait.* The blue whale (*Balaenoptera musculus*) and the fin whale (*Balaenoptera physalus*) are each other's closest relatives, and there is evidence that the two species are able to interbreed (see Figure 8-3).[15] In our view, however, the blue whale adds a great deal to our planet's overall biodiversity as the largest animal ever to live on the Earth. The methods described here would not place a high priority on the blue whale; indeed, they would give it a low priority, given its close relationship with the fin whale. Species that are outstanding in one or a few characters, but that have several close relatives, generally receive low priority in systems such as those of Williams et al.

Systematist Daniel P. Faith has produced selection procedures for setting priorities that bear a close resemblance to those of the British Museum group. Although important differences exist between Faith's work and the British Museum proposals, both assume the existence of a well-worked-out phylo-

---

*This is a result of adhering to strict cladistic techniques, which weight all characters equally in determining evolutionary relationships.

geny (that is, an explicit understanding of the evolutionary relationships among species). Faith's procedure, in contrast to the methods of the British Museum group, attempts to assess biodiversity by focusing on the characteristics and features of species. "When biodiversity is defined at this level," he writes, "the oft-stated conservation goal of 'protecting biodiversity' translates into protecting as much of this feature-diversity as possible."[16] Like the British Museum group, Faith's proposal focuses on a single level, that of features and characteristics of species, and does not allow for other levels of biodiversity to be incorporated into his method. In addition, he assumes that all features have equal worth, since scientists cannot today know how they might make use of any given feature in the future. However, biologists actually do have some indication that certain features used to develop phylogenies will be more useful than others; the patterns of veins in leaves, the forms of flowers, and the structures of chemical compounds may be important features in constructing a phylogeny for a group of plants, but it is unlikely that leaf venation or flower form will ever prove economically important, whereas certain chemical compounds may.

Martin Weitzman, an economist at Harvard, has produced a generalized measure of diversity to identify which members of a group of related species (a clade) add the most diversity to the group. The crane example that we used to explain cladograms on page 97 is borrowed from Weitzman's work.[17] His method can also identify those species whose extinction would be the greatest loss to the group's diversity, according to his definition. Weitzman employed genetic distance to assess diversity of species in the crane family, although his method is specifically tailored to using other measures of difference. In refinements of his basic model he takes further steps, incorporating information on the probable marginal costs of conservation of each species to help set priorities.

Regardless of whether the elements of biodiversity being considered are species or features of species, and whether economic considerations are included or not, the underlying goal of all these methods is to protect the most diverse subset of biodiversity possible, with each method defining *diverse* in its own manner. We believe that all of these methods offer potentially useful perspectives on the problem of assessing biodiversity, but that by trying to capture the *single* best measure of the diversity in biodiversity, they necessarily fall short. Given the many levels of biodiversity, and the many ways in which elements of biodiversity differ from each other, we believe that no one measure of biodiversity alone will help conservation organizations as they articulate and set priorities for conservation.

Protecting the *Most Valuable* Biodiversity

The three goals discussed so far all suffer from significant flaws. Attempts to "protect all" run the risk of expending limited resources on a few threatened elements of biodiversity while at the same time allowing the more valued elements to become threatened. Measures that would enable one to decide which assemblages had the "most" biodiversity and which the "least" cannot capture the depth of the term *biodiversity* and thus will fail to protect what we really want to protect. Finally, given that each actor has his or her own definition of "diversity," as we discussed in Chapter 4, attempts to identify the "most diverse subset of biodiversity" by using a single measure are bound to fail.

We recommend that conservation organizations articulate what they find to be valuable among elements of biodiversity and attempt to protect these elements. Wherever possible, the priority-setting organization should consult all of the individuals and organizations that stand to gain or lose by these actions—the stakeholders—to determine what they value regarding biodiversity. Once the stakeholders' values have been articulated, the organization can develop selection procedures that embody those values.

This goal is motivated by two facts: 1) the human race does not have sufficient resources to protect all of biodiversity, and 2) our species has already accumulated a great deal of useful knowledge about different elements of biodiversity (even though we have a great deal more to learn). Furthermore, this knowledge is located neither in a single culture nor in a single discipline, such as academic biology. Individuals each have their own sets of values and preferences, just as organizations each embody certain values and have interests. Articulating what is valuable in biodiversity is a political process that depends on the accumulation of many stakeholders' views. We do not suggest that conservation organizations hold biodiversity popularity contests. Rather, because we are attempting to understand what it is in biodiversity that humans find valuable, we must include as wide a range of human voices as possible. Just as the plant systematist can highlight the importance of a relict species, the pharmacologist can indicate the potential value of a class of natural compounds, and the local inhabitant can point out a habitat of great spiritual value or a species of symbolic importance. No single individual or organization can express the full range of values that humans hold toward elements of biodiversity, or identify all types of worth found in the biodiversity of a region. For example, in the Biodiversity Support Program Workshop described at the start of this chapter, the fourth group articulated that it valued intact ecosystems, a value that the other groups had not expressed.

To a large degree, the other central goals exclude most stakeholders' voices. The goal of "protecting all" is so absolute in nature as to exclude the nuances of human preferences and interests. Although not absolute, the goals of protecting the "most" and protecting the "most diverse" have the appearance of scientific objectivity and are typically described as being so technical that only scientists are allowed to judge these matters. Moreover, proponents of these goals frequently act as if there were just a single right method for determining the "most" or "most diverse," and assume that a major task of scientists is to uncover this method.

We hope to see the goal of protecting the most valuable put into practice so that it gives voice to a wide variety of people, which should help create a better fit between conservation goals and conservation actions than might otherwise be achieved. In the next section we discuss the ways in which central goals influence the entire priority-setting process, from inventories to action plans.

THE PRIORITY-SETTING PROCESS

Most conservation organizations embody certain values and goals that shape the way they define their task; examples include protecting rare species or ecosystems, preserving local open space, or managing specific game species. Even when different organizations share the goal of "protecting biodiversity," they may in practice define their task in very different ways. Where one organization may emphasize locally rare entities, another might place top priority on globally rare ones or economically important strains of domestic plants.

Priority-setting processes are shaped by the values, goals, and resources of their users (see Figure 8-4). Ideally, the entire process functions as an accurate reflection of the organizations' interests, and both helps them to set priorities among the different elements of biodiversity being considered for protection and allows them to act on their priorities in an effective manner. An organization's resources (or lack thereof), along with its values, affect the entire priority-setting process. If, for example, an organization would like to protect a certain endangered species, but lacks the technical expertise or financial resources needed to establish a captive breeding program or to purchase extensive habitat, then the species will not become a high-priority project for the organization. For nearly 15 years, the Massachusetts Natural Heritage and Endangered Species Program had the Hyannis Ponds complex near the top of its list of biodiversity sites in the state needing protection, but the program lacked the funds to take any action on the site; it was only after a bond bill was passed by the state legislature that resources became available to move the site from the wish list to the action list (see Chapters 3 and 7 for more information on these ponds).[18]

SCALE OF CONCERN

Throughout the entire priority-setting process certain aspects of the Earth's biodiversity are considered relative to some scale of concern. Geographic scale is perhaps the easiest to imagine, but the process may also focus on parts of the taxonomic and temporal "landscapes" (see Chapter 5). An organization may have a global, national, or local scale of concern, and may ignore everything (or nearly everything) outside the region of concern. The Endangered Species Act, for example, takes the United States as its primary geographic scale of concern, although the act also calls for action on internationally threatened species, such as those listed by the Convention on International Trade in Endangered Species. Some organizations have a restricted taxonomic focus, as do BirdLife International (formerly the International Council for Bird Preservation) and the Xerces Society (which is dedicated to the conservation of invertebrates), while others, such as the New England Wild Flower Society, have restricted geographic and taxonomic scales of concern.

An organization may also restrict its focus to a certain portion of the temporal landscape for evaluating conservation decisions. For example, Norton and Ulanowicz suggest that a time period of about 150 years is most appropriate for conservation decisions, whereas some discussions of minimum viable populations take a 1000-year horizon.[19] In practice, many conservation organizations are under pressure to make decisions quickly, in the face of rapid changes. Mark Primack, director of the Wildlands Trust of

Southeastern Massachusetts, a local land trust, has discussed with our class that his organization is having to make decisions on a two- to five-year horizon.[20] The town of Plymouth, home to a number of locally rare species and globally rare ecosystems, has several new railway terminals opening, which will lead to significant new residential and commercial development—and this growth is happening on top of a nearly 100% population increase during the 1980s. The next few years will present opportunities to conserve wild areas in Plymouth; after that, development will have proceeded so far that many conservation options will no longer be available.

In sum, when a conservation organization selects a specific scale of concern within which to focus its efforts, it should be aware of the biases that each scale produces. Special care should be taken to consult with partner organizations employing different scales of concern to make sure that conservation efforts are neither excessively duplicated nor misspent by selecting a given scale of concern.

VOICES

In April 1992, the Biodiversity Support Program sponsored a Conservation Needs Assessment workshop in Papua New Guinea. The workshop included "representatives of the government of Papua New Guinea, USAID [U.S. Agency for International Development], numerous scientific and research institutions and museums, social scientists and legal scholars, NGOs [nongovernmental organizations], and local Landowners' groups."[21] A report written about the workshop process itself describes how the participants' different backgrounds and views made for a wide-ranging and often heated debate.[22] Although these differences led to tension, we believe that they added to the working groups' conception of what it is they should be trying to conserve in Papua New Guinea. The consensus recommendations of the workshop participants called for the development of a National Environment and Conservation Plan and a Natural Resources Option Centre to complement the extensive biodiversity maps produced during the workshop.

The range of participants in this workshop was representative of a wide range of values about biodiversity in Papua New Guinea. We hope that this inclusive format will be a model for future conservation decision making. What we are advocating hearkens back to our comparison in Chapter 1 between the concepts of biodiversity, rights, and equality. Rights and equality are the cornerstones of democracy, for in a democracy all citizens have the right to have their voices heard. The beauty of the democratic system is that, more than any other system of government, it honors diversity, the diversity of views of all of its citizens. Our observations have been that the citizens of the world value nature in diverse ways, and the more we listen to this diversity of voices, the better we will understand what to conserve.

Much of this chapter described answers to the question, What biodiversity should we protect? By including so varied a group of participants, the Biodiversity Support Program workshop generated a large number of different answers to this question. In our final chapter we turn to a related query, one that we have been building toward answering for the length of the book:

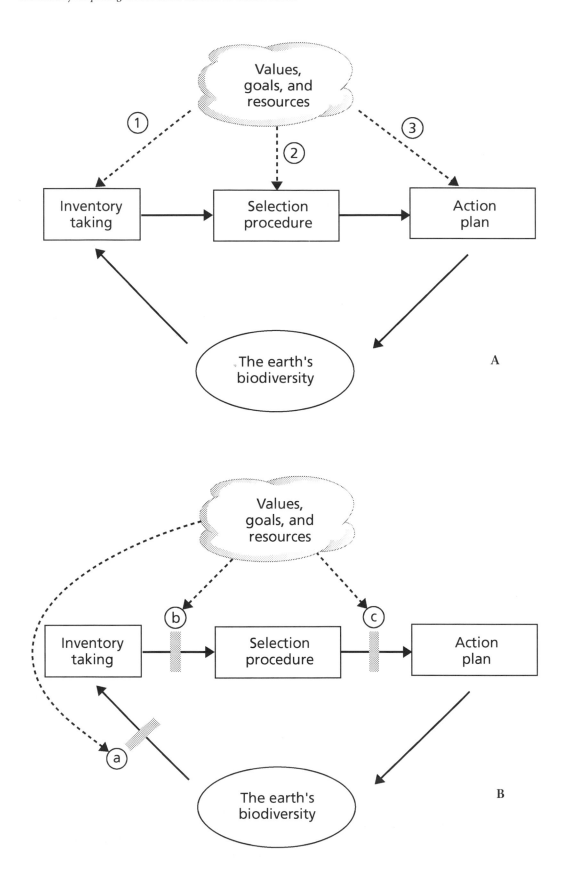

Figure 8-4.

Values, goals, and resources affect the entire priority-setting process. Here are two schematic views of how values, goals, and resources shape and constrain the entire priority-setting process. These factors influence the process in two main ways: through the selection of specific inventory techniques, methods for setting priorities (selection procedures) and action plans; and by filtering the flow of data and options throughout the process. Note that the process incorporates a feedback loop—action plans affect elements of biodiversity, creating a different situation for the next inventory. Heavy arrows indicate the flow of information and actions concerned with biodiversity protection.

(A) Values, goals and resources affect the selection of key components.
   The dotted lines labeled 1) through 3) indicate the importance of values, goals, and resources in the selection and shaping of inventories, methods, and action plans.

   1)   Values, goals, and resource limitation affect the choice of inventory techniques, which affects the information available for organizations. For instance, did the organization call for an inventory of only woody plants, or did it also include some groups of invertebrates? Did the organization emphasize gathering degree-of-threat information or taxonomic status? Was the scope of the inventory similar to the proposed five-year, $100-million All Taxa Biodiversity Inventory in Costa Rica, or did it consist of a pair of graduate students recording the plants and birds present at a site over the course of a year?

   2)   The specific choice of selection procedure makes a big difference in the outcome of the process. The organization's central goals greatly influence the choice of procedure, and each procedure assesses biodiversity using different algorithms and different types of data. Selection procedures that focus on degree of threat, number of species present, and taxonomic spread of species within a group of related species will all produce different priorities for a single site.

   3)   The range of action plans employed by an organization is limited by its values, goals, and resources, thereby constraining the outcomes of the priority-setting process. For example, local land trusts protect elements of biodiversity by acquiring land or conservation easements on land but do not embark on captive breeding programs. Large zoos and arboreta, on the other hand, may undertake captive breeding programs but rarely purchase land for in situ conservation.

(B) "Filters" are created by values, goals, and resources.
   The gray bars labeled a) through c) represent "filters" of information and options, filters that stem from the values, goals, and resources of the organization setting priorities. The practical definition of biodiversity chosen by the organization plays an important filtering role as well. If the definition is species-based, information about genes, ecosystems, and other levels of biodiversity will be filtered out from the start of the process.

   a)   As we discussed in Chapter 7, inventories do not attempt complete biodiversity surveys; from the very start they focus on only certain groups of species (e.g., seed-bearing plants and vertebrates) or levels of biodiversity (e.g., species and ecosystems rather than genes). Thus, most of the information about the biodiversity of a site gets filtered out even *before* an inventory is created, largely as a result of the practical definition of biodiversity chosen by the organization. Elements not included in the inventory will not be considered later during the priority-setting process.

   b)   After the inventory is completed, some of the biodiversity elements listed in the inventory may be left out of the priority-setting process. For example, an organization might use an already-existing inventory that covered many taxonomic groups, some of which are of little interest to the organization. In such cases, these inventory data are simply ignored. Furthermore, depending on the goals of the organization, information can be added to the inventory data. Such additions may include the following about species: their status (rare or endangered?), their biology (widespread or restricted in distribution; do they play a special role in the ecosystem?), their taxonomic position (many close relatives or taxonomically isolated?), and their importance to humans (important economically, symbolically, or spiritually?).

   c)   Elements of biodiversity that are being considered for protection may be selected or excluded for clearly articulated reasons, or the process may be one of default. For example, BirdLife International has clearly stated that its interests lie in protecting endangered bird species. In contrast, no species of bacteria are listed as endangered under the Endangered Species Act, largely because we do not have the knowledge to determine whether a bacterial species is threatened or not. Even though a certain element of biodiversity has been given a high priority and proposed for protection, the organization may not put an action plan into place for the element due to financial or political constraints.

Why should we protect biodiversity? We do not offer a single answer that settles the issue; rather, we offer several, and suggest that each conservation decision maker, each person who cares about the biodiversity legacy our generation leaves behind, should attempt to develop his or her own answers.

References

1. Biodiversity Support Program, Conservation International, The Nature Conservancy, Wildlife Conservation Society, World Resources Institute, and World Wildlife Fund, *A Regional Analysis of Geographic Priorities for Biodiversity Conservation in Latin America and the Caribbean* (Washington, D.C.: Biodiversity Support Program, 1995), 7.

2. Biodiversity Support Program, Conservation International, The Nature Conservancy, Wildlife Conservation Society, World Resources Institute, and World Wildlife Fund, *A Regional Analysis of Geographic Priorities for Biodiversity Conservation in Latin America and the Caribbean* (Washington, D.C.: Biodiversity Support Program, 1995), 13–14.

3. Biodiversity Support Program, Conservation International, The Nature Conservancy, Wildlife Conservation Society, World Resources Institute, and World Wildlife Fund, *A Regional Analysis of Geographic Priorities for Biodiversity Conservation in Latin America and the Caribbean* (Washington, D.C.: Biodiversity Support Program, 1995), 27.

4. Biodiversity Support Program, Conservation International, The Nature Conservancy, Wildlife Conservation Society, World Resources Institute, and World Wildlife Fund, *A Regional Analysis of Geographic Priorities for Biodiversity Conservation in Latin America and the Caribbean* (Washington, D.C.: Biodiversity Support Program, 1995), 28.

5. Convention on International Trade in Endangered Species of Wild Fauna and Flora, Article II (1).

6. See U.S. Government Accounting Office, *Endangered Species Act: Types and Number of Implementing Action* (GAO/RCED-92-131BR, 1992), 26, 36–38.

7. Bryan G. Norton, "On What We Should Save: The Role of Culture in Determining Conservation Targets," in *Systematics and Conservation Evaluation*, ed. P. L. Forey, C. J. Humphries, and R. I. Vane-Wright (Oxford: Oxford University Press, 1994), 25–26.

8. J. M. Scott, B. Csoti, J. D. Jacobs, and J. E. Estes, "Species Richness," *Bio Science* 37 (1987):782–788.

9. Megadiversity countries: Russell A. Mittermeier, "Primate Diversity and the Tropical Forest," in *Biodiversity*, ed. E. O. Wilson (Washington, D.C.: National Academy Press, 1988); World Conservation Monitoring Centre, *Global Biodiversity: Status of the Earth's Living Resources* (London: Chapman & Hall, 1992); World Resources Institute, *World Resources 1996–97* (New York: Oxford University Press, 1996).

10. World Conservation Monitoring Centre, *Global Biodiversity* (London: Chapman & Hall, 1992), 42.

11. R. L. Pressey, C. J. Humphries, C. R. Margules, R. I. Vane-Wright, and P. H. Williams, "Beyond Opportunisms: Key Principles for Systematic Reserve Selection," *Trends in Ecology and Evolution* 8 (1993): 124–128.

12. R. L. Pressey, C. J. Humphries, C. R. Margules, R. I. Vane-Wright, and P. H. Williams, "Beyond Opportunisms: Key Principles for Systematic Reserve Selection," *Trends in Ecology and Evolution* 8 (1993): 125.

13. P. H. Williams, R. I. Vane-Wright, and C. J. Humphries, "Measuring Biodiversity for Choosing Conservation Areas," in *Hymenoptera and Biodiversity*, ed. J. LaSalle and I. O. Gauld (Wallingford, United Kingdom: CAB International, 1993), 310.

14. P. H. Williams, R. I. Vane-Wright, and C. J. Humphries, "Measuring Biodiversity for Choosing Conservation Areas," in *Hymenoptera and Biodiversity*, ed. J. LaSalle and I. O. Gauld (Wallingford, United Kingdom: CAB International, 1993), 310–311.

15. R. Spilliaert, G. Vikingsson, H. Arnason, A. Palsdottir, J. Sigurjonsson, and A. Arnason, "Species Hybridization Between a Female Blue Whale (*Balaenoptera musculus*) and a Male Fin Whale (*B. physalus*): Molecular and Morphological Documentation," *Journal of Heredity* 82, no. 6 (1991): 269.

16. Daniel P. Faith, "Phylogenetic Pattern and the Quantification of Organismal Biodiversity," *Phil. Trans. R. Soc. Lond. B* 345 (1994): 46.

17. Martin Weitzman, "What to Preserve: An Application of Diversity Theory to Crane Conservation," *Quarterly Journal of Economics* 108, no. 1 (1993): 157–184.

18. Henry Woolsey, director of Massachusetts Natural Heritage and Endangered Species Program, personal communications, 1994–1996.

19. Bryan Norton and Robert Ulanowicz, "Scale and Biodiversity Policy: A Hierarchical Approach," *Ambio* 21 (1992): 244–249; Mark L. Shaffer, "Minimum Population Sizes for Species Conservation," *Bioscience* 31 (1981): 131–134.

20. Mark Primack, personal communication, October 1995.

21. J. Fred Swartzendruber, "Executive Summary," in *Papua New Guinea Conservation Needs Assessment Volume I*, ed. Janis B. Alcorn (Washington, D.C.: The Biodiversity Support Program, 1993), xi.

22. Janis B. Alcorn, "The CNA Workshop," in *Papua New Guinea Conservation Needs Assessment Volume I*, ed. Janis B. Alcorn (Washington, D.C.: The Biodiversity Support Program, 1993).

9 *Future Directions*

All nature is so full, that that
district produces the greatest
variety which is most examined.

Gilbert White,
The Natural History and Antiquities
of Selborne (1768)[1]

THE BILTMORE STICK

When we take our students to northern Vermont, we introduce them to the conservation problems and land use issues of the region, including the massive impacts of paper companies and vacation home developments. We also expose them to elements of biodiversity that most of them have never seen before, such as old-growth hardwood forests and quaking bogs. One of the highlights of the annual trek is our meeting with Brendan Whittaker of the Vermont Natural Resources Council, a forester, pastor, farmer, and former Secretary of the Environment for Vermont. We spend the day walking with Bren through his 63 acres of forest. He points out sections of his land that he has left untouched, in order to let nature take its course, and other sections that he has weeded, thinned, planted, nurtured, and cut according to various forest management strategies of the last few decades. He speaks to us not only of his land but also of the plight of the rural poor, and of their aversion to having values imposed upon them from without—from environmentalists who were born and raised in urban areas, most of whom have never thought of what it means to live off the land. Along with the insights and knowledge that he shares with the students, he introduces them to a simple, yet surprisingly sophisticated, measuring implement—the Biltmore stick (see Figure 9-1).

At first glance, the Biltmore stick appears to be a handmade wooden ruler, and in fact, Bren made his stick during his student days at forestry school. A closer look reveals that, as a ruler, the stick is inaccurate—although it is marked off in inches, the marks are not evenly spaced (see Figure 9-2). Each additional "inch" is smaller than the previous "inch"; as one proceeds from the zero mark to the opposite end of the stick, the "inch" marks get closer together. As we stand among the red pines that he planted some 35 years ago to provide for his young children's future, Bren explains to a volunteer from our class how to use this odd device: hold your arm slightly crooked, with the stick resting on the side of a tree whose diameter you want to measure.

Figure 9-1.
Using the Biltmore stick to determine the diameter of a tree. Each forester had to build a customized Biltmore stick. (Photograph by Solomon Perlman, Dan Perlman's grandfather, and a forester in the early twentieth century)

Close one eye, line up the zero mark along one edge of the tree, and read off the diameter by sighting the other edge of the tree. When we measure the tree with a tape measure to confirm the readings that Bren and the volunteer made with the Biltmore stick, we invariably find that they are quite close to the true diameter—though Bren's estimate is always a little closer to the mark than the student's. This discrepancy arises because each forestry student makes his or her own Biltmore stick, as each student has a slightly different arm length and accordingly stands at a slightly different distance from the tree; in other words, each forester would have a slightly different perspective on exactly the same tree. To assess the diameters of trees accurately, each forester needs a customized device.

As we have discussed throughout this book, we believe that conservation decision makers should create customized "devices" to help them assess biodiversity and set priorities for its protection, much as each forester created a customized Biltmore stick. Granted, foresters developed the Biltmore stick to measure a single, well-defined variable—the diameter of trees—while conservation decision makers are developing methods to assess biodiversity, an entity that is complex and whose definition changes from person to person. The parallels, however, are many. In both cases, the entity being assessed (standing trees and biodiversity) has great potential worth to humans. This worth derives from the ability of trees and biodiversity to satisfy

Figure 9-2.
Close-up of Bren Whittaker's Biltmore stick. Note how the "inch" measurements are not spaced equally; this variation is the adjustment that allows the user to measure a tree's diameter accurately from a single vantage point. (Photograph by Dan L. Perlman)

so many deep human values—the desire for shelter and warmth, food, medicines, spiritual renewal, and aesthetic experiences, to name a few.

Overall, the goal of protecting biodiversity is shared by many. But different conservation decision makers bring different perspectives to the goal of protecting biodiversity, just as different foresters bring different perspectives when viewing a given tree. To best protect biodiversity for future generations, our generation needs to identify as many different values as possible that humans hold toward biodiversity, and to build a series of "biodiversity Biltmore sticks" to assess the many different conceptions of biodiversity. In this chapter we discuss different types of values that humans hold toward biodiversity. We do not intend this list to be exhaustive; rather, we hope that it will be seen as a starting point, and that others will add to it.

We do not believe that a single, all-purpose "measure" of biodiversity exists, nor do we believe that one should be developed. Not only are the many elements of life with which we share our planet diverse, but the ways in which human beings value these elements are diverse. No single metric could ever include all of the values that humans hold toward biodiversity. As philosopher Holmes Rolston III has written, "The extent of biodiversity is not an easy question to answer, either in a technical, scientific or in a philosophical . . . sense. The difficulty rises from its richness, from the diversity of diversity, if you will."[2] How, then, are we to support the efforts of conservationists as they decide how to set priorities using scarce conservation resources? What guidance do we offer the decision maker who attempts to protect our biodiversity legacy?

We believe that the very diversity of human values and responses toward biodiversity holds the key to selecting the best criteria for biodiversity protection. We have come to see the question of priority setting for biodiversity conservation as a political process in which the voices, values, ideals, and interests of as wide a variety of people as possible need to be considered. In large measure, biodiversity conservation has often been treated as a biological issue—a question for biologists to debate, offer counsel about, and act upon. Starting in the 1980s, however, with the publication of the *World Conservation Strategy* and *Our Common Future,* and reaching the level of an international treaty in 1992 with the Convention on Biological Diversity, more conservationists began to recognize the loss of biodiversity as an issue affecting all of humanity and not just biologists and nature lovers.[3]

Is it feasible and reasonable for a wide variety of people to attempt to set priorities for biodiversity conservation? We believe it is, for it is this debate among many voices that is the essence of democracy. In a democratic society, no single individual and no single group of professionals sets the priorities for the spending of limited resources on issues of public good. While educators advise lawmakers and the public on spending for schools, and physicians, public safety officials, and the military advise on their respective fields, none of these professionals makes the decisions about apportioning resources among these different public needs. Instead, the differing needs are all weighed against each other in the arena of public policy.

In the biodiversity realm, no single measure can decide for us the relative worth of protecting the endangered California condor versus the worth of protecting the Hyannis Ponds on Cape Cod. Nor can any measure decide for us whether improving preschool education is worth more than improving cancer survival rates or upgrading the protection given the Hyannis Ponds. Yet each of these activities has its strong advocates in public life, each requires the input of money and human resources, and it may be that not all can be fully funded.

The reality of our world today is that even in the wealthiest societies, many worthy projects are competing for scarce public funds. Yet depending on how large a program conservationists propose, the protection of biodiversity may be one of the most expensive public programs ever carried out. It is certainly, in our view, among the most important. We believe that under conditions of limited resources, the wisest goal for protecting biodiversity is to protect the most valuable aspects of biodiversity, with "most valuable" being defined by the inputs of as many stakeholders as possible, as we discussed in Chapter 8. With more voices, we increase the likelihood that elements of biodiversity that are important to different segments of human society, and elements that are important in their own right, will receive attention and protection.

Clearly, including many stakeholders in the process of selecting criteria for biodiversity conservation will take time and effort. The process may even lead to occasional lost opportunities, if it is especially slow, and will at times lead to priorities quite different from those that professional conservation biologists might have recommended. But we must recognize that our planet's biodiversity is not the sole domain of conservation biologists, that the biodiversity legacy belongs to all humans, and as such, all should have a say in shaping it—with the help and guidance of professionals. As conservation biologist Michael Usher suggests, "Values, then, are determined by society, though it is often true that the scientist suggests to the society what the values should be."[4]

To date, much of the emphasis in conservation biology has been on the scientific aspects of the discipline; issues such as nature reserve design and the problems of small populations have received the bulk of attention. Furthermore, it appears that many conservation biologists view the assessment of biodiversity and the priority-setting process as technical issues. We suggest that the time is now ripe for the discipline of conservation biology to focus on the art and politics of biodiversity assessment and priority setting. The realm of public policy is where these issues will be fought out, where they will live and die. The efforts of our generation will be judged, centuries from now, not by the answers we offer to questions of how to design nature reserves, but by the fundamental questions we choose to ask in the first place.

RATIONALES FOR PROTECTING BIODIVERSITY

In Chapter 3 we discussed the different values that humans hold and the types of worth that they attach to elements of biodiversity. To best satisfy the needs of humankind today and in the future, as well as safeguarding the needs of the other living creatures of our planet, we need to consider the values and interests of a wide variety of individuals and cultures. As one way of doing so, we discuss different rationales for protecting biodiversity that we have sifted from the literature and practice of conservation biology.

Just as biodiversity can be described on many different levels, so can the values that humans bring to biodiversity and the worth that they find in biodiversity. In some of what follows, we describe very specific elements of biodiversity, such as genes and natural chemicals found in wild organisms, while at other times we describe elements that are so broad as to be considered a general description of much of nature, as when people discuss their desire for wild places. Some rationales are intended to make a case for saving biodiversity generally, while others are essentially decision rules for priority setting, indicating how to select different assemblages of biodiversity for protection.

We intend the following list to be used as a starting point, to help others think about their own rationales for protecting biodiversity. We have chosen not to judge the relative merits of these different rationales; instead, we are putting them in play to demonstrate the breadth of both the values that humans bring to this issue and the worth that our species has discovered in biodiversity. Along with each rationale we list the values and categories of worth that support the rationale. Chapter 3 contains an extensive discussion of the different types of values and worth used here.

Existing elements of biodiversity have a right to continued existence and deserve protection

Values: moralistic and humanistic (Kellert),* religious, the "Noah principle"[5] (Ehrenfeld)

Worth: existence value, bequest value

Notes: As Ehrenfeld points out, this rationale for protecting biodiversity has a long history in Western religions, and much of the focus on it stems from the religious and spiritual realms (see Figure 9-3). This rationale also includes arguments based on the intrinsic rights of nonhuman individuals and species.

Figure 9-3.
Noah's Ark. The ark protected all species and enabled them to survive the biblical Flood. Today, the ark stands as an important symbol of the tasks facing conservationists.

*These values are from Stephen Kellert's work; see Chapter 3 for a discussion of them.

Figure 9-4.
The dodo. Perhaps the most recognizable symbol of human-caused extinctions, this bird stands as a reminder of what may happen to the threatened species of today.

Threatened elements of biodiversity deserve protection

Values:	naturalistic, ecologistic-scientific, symbolic, humanistic, moralistic (Kellert), protect all (Chapter 8)
Worth:	"keep every cog and wheel" (Leopold),[6] option value, existence value
Notes:	The Lacey Act of 1900 was the first federal wildlife protection act in the United States, and the Endangered Species Act of 1973 is the most powerful; these are both based on this rationale. Extirpated elements of biodiversity, such as the dodo, can be powerful symbols representing this rationale (see Figure 9-4).
Allied rationales:	"Rare elements of biodiversity deserve protection." "Endemic elements of biodiversity deserve protection." Rarity and endemism can both be viewed as independent reasons for protecting an element or as the first step toward being threatened, since rare and endemic elements run a higher risk of being extirpated than common and widespread elements. "Common and weedy elements of biodiversity deserve little protection." This is the converse of the previous rationales and is often an unstated undercurrent in conservation decisions.

Ecosystems and species that provide important commodities deserve protection (e.g., timber, fish, raw material for industrial processes)

Values: utilitarian (Kellert)

Worth: direct use value

Notes: Many ancient societies have long-standing traditions of areas that do not get hunted or where trees are not harvested; the result of these actions is to protect a breeding population or seed source for economically important species. During medieval times, hunting preserves were established to retain large local populations of game animals for the pleasure of royalty. In modern societies, restrictions on logging and fishing derive from the same motive (see Figure 9-5). Many raw materials for chemical, industrial, and pharmaceutical processes are taken from domesticated and wild species such as oil palms, corn, and soybeans.

Figure 9-5.

Commercial fishing boats. Boats like this harvest important commodities from nature. Today, however, overfishing throughout the world has led to a need for strict protection of many fishing grounds and commercially important species.

Figure 9-6.
Salt marsh on Martha's Vineyard. Wetlands, such as this marsh, help to filter water and serve as important breeding grounds for many aquatic organisms. The public is becoming aware of the important services that ecosystems provide for humanity. (Photograph by Dan L. Perlman)

Ecosystems deserve protection if they provide important services (e.g., carbon sequestration, watershed protection, flood control, filtration)

Values:	Utilitarian (Kellert)
Worth:	indirect use value
Notes:	Watershed protection has a long history, and the roles that intact ecosystems play in other regards are becoming better understood and more widely recognized (see Figure 9-6). Forests and prairies play important roles in storing carbon, which helps to moderate the global climate. Our species is now begining to understand that we depend on ecosystem services for our survival.

Figure 9-7.

The Eurasian crane (*Grus grus*). The ancient Greeks watched this species migrate overhead and nest in their land. Like other crane species the world over, these birds have inspired powerful emotions in those who view them. Today, the species no longer nests in Greece. (See Color Plate 15)

Charismatic, beautiful, and symbolically important species deserve protection

Values:	naturalistic, aesthetic, symbolic (Kellert)
Worth:	existence value
Notes:	Attractive, interesting, and important animal and plant species have long been important in human life. Detailed and beautiful cave paintings of large animals, dating from at least 30,000 years ago, are among the earliest examples of art, and most human societies either keep animals as pets or worship certain animal species. E. O. Wilson has proposed the biophilia hypothesis to explain the human attraction to (and repulsion from) certain types of organisms; he suggests that biophilia is a genetic predisposition of our species.[7] As conservation organizations recognize, this rationale works very well with the general public. Species that remind us of ourselves (primates, social mammals, and birds such as cranes) are also covered by this rationale (see Figure 9-7).

Figure 9-8.
Intact ecosystems have special appeal. Ecosystems with integrity, such as the prairie depicted here, play a role as flagship sites in efforts to protect biodiversity.

Ecosystems that have integrity, and elements of biodiversity that add to the integrity of ecosystems, deserve protection

Values:	ecologistic-scientific, aesthetic, naturalistic (Kellert)
Worth:	indirect use value, existence value
Notes:	"Pristine" habitats, such as primary forests, have long received special attention from conservationists (we acknowledge that most terrestrial habitats have been affected by humans in some way, and doubt that many habitats are truly pristine). Discussions about the integrity of ecosystems are beginning to appear in the literature, and we expect that this trend will continue (see Figure 9-8).[8]
Allied rationales:	"Keystone species in ecosystems deserve protection." "Native elements in ecosystems deserve protection."[9] Elements within ecosystems that add to the systems' integrity are especially deserving of protection. "Exotic elements of biodiversity invading ecosystems should be eradicated." This is the converse of the previous rationales, and is often an unstated undercurrent in conservation decisions.

Figure 9-9.
A living fossil. The coelacanth (*Latimeria chalumnae*) was known only from fossils some 70 million years old—until its rediscovery in 1938 off the African coast. It is one of the best-known relict species, organisms that are especially treasured by biologists. (For a description, see Peter B. Moyle and Joseph J. Cech Jr., *Fishes*: *An Introduction to Ichthyology,* 3d ed. [Englewood Cliffs, New Jersey: Prentice Hall, 1996].)

Species that are taxonomically, morphologically, behaviorally, or physiologically distinct deserve protection

Values:	ecologistic-scientific, utilitarian (Kellert)
Worth:	option value, existence value
Notes:	Biologists have long treasured relict and oddball species for what they tell us about evolution and for the sheer joy of contemplating the wonders of nature (see Figure 9-9). In addition, biologists have proposed that taxonomically distinct species may hold genetic information that can prove important for sustainable development by coding for important natural products that can be used in the pharmaceutical and chemical industries.
Allied rationales:	"Elements of biodiversity that evolve quickly or slowly or are especially evolutionarily successful deserve protection."
	Under this rationale, it is not the features of the organism that stand out but rather the species' rate of evolution or evolutionary impact.
	"Taxa that are close relatives of domesticated species deserve protection."
	This rationale is based on the recognition that such taxa may offer domestic crops genetic resistance to pests and diseases.

Figure 9-10.
Coral reef. Coral reefs, tropical forests, and other sites rich in biodiversity often have powerful impacts on people visiting them.

Ecosystems containing many species deserve protection

Values:	ecologistic-scientific, aesthetic, naturalistic (Kellert)
Worth:	option value, existence value
Notes:	This rationale stems from two very different traditions. The first is essentially a bet-hedging strategy, trying to save the most biodiversity per amount of effort expended. The second, an aesthetic argument, grows out of the great impact that diverse natural settings can have on humans (see Figure 9-10).

Figure 9-11.
Bosque Eterno de los Niños (The Children's Eternal Forest). This forest preserve in Costa Rica was created with funds raised by children around the world. (Photograph by Dan L. Perlman)

Nearby elements of biodiversity deserve protection more than distant elements

Values: Protect the most valuable biodiversity
Worth: Direct and indirect use values, option value
Notes: This rationale is widespread; individuals and organizations often undertake major efforts on behalf of local elements of biodiversity that they would not perform for distant ones. For example, in 1995 the City of Huntsville, Alabama, along with The Huntsville Land Trust and Madison County, purchased approximately 223 hectares of land for about $5 million. While the site hosts a species proposed for listing (*Clematis morefieldii*, a perennial vine), the facts that the land was beautiful and offered recreational and educational opportunities within the city limits were the key issues in deciding to purchase the land.[10]

Allied rationale: "Distant elements of biodiversity deserve protection more than nearby elements" (see Figure 9-11).
This rationale is the converse of the one above; people often find the faraway more alluring than the local, which can appear mundane.

Like many others, we believe that today humankind faces an enormous task and a great responsibility: if we do not carefully and quickly act to protect biodiversity, our world and the world we leave to our children will be far poorer. We can no longer claim ignorance, or assume that the planet's resources are virtually infinite, as previous generations might have. Like the scholars and scribes of Alexandria and Byzantium, we now know that the actions we take—and do not take—will reverberate through the rest of time; the literature that they did not protect, and the species and other elements of biodiversity that we do not protect, are lost forever.

Each year when our class reaches the topic of extinction, we read about the passenger pigeon, whose flocks would darken the skies for hours as they passed overhead. We find ourselves considering the biodiversity legacy that was left to our time and the holes in its fabric, and we ponder the legacy that we are leaving for generations to come. Will humans in the future be able to travel through great temperate and tropical forests, depend on food stocks secured by the genetic diversity of wild relatives, and turn to nature for spiritual renewal? Or will we leave them a world that supplies only their barest needs, with little security and inspiration to be found in nature? What will they imagine, a century from now, when they read the epigraphs from the beginning of this book? Will they know the crane and bluebird as living creatures, or merely as illustrations in books and specimens in museums? Those are the questions that we must answer for the generations to come.

References

1. Gilbert White, *The Natural History and Antiquities of Selbourne* (Harmondsworth, England: Penguin Books, 1977).
2. Holmes Rolston, *Conserving Natural Value* (New York: Columbia University Press, 1994), 35–36.
3. IUCN, *World Conservation Strategy* (Gland, Switzerland: IUCN, 1980); IUCN, UNEP, WWF, and World Commission on Environment and Development, *Our Common Future* (Oxford: Oxford University Press, 1987).
4. Michael B. Usher, "Wildlife Conservation Evaluation: Attributes, Criteria and Values," in *Wildlife Conservation Evaluation,* ed. Michael B. Usher (London: Chapman & Hall, 1986), 8.
5. David Ehrenfeld, *The Arrogance of Humanism* (Oxford: Oxford University Press, 1978).
6. Aldo Leopold, *Round River* (New York: Oxford University Press, 1993).
7. Edward O. Wilson, *Biophilia* (Cambridge, Massachusetts: Harvard University Press, 1984); Stephen R. Kellert and Edward O. Wilson, *The Biophilia Hypothesis* (Washington, D.C.: Island Press, 1993).
8. Robert Costanza, Bryan G. Norton, and Benjamin D. Haskell, *Ecosystem Health* (Washington, D.C.: Island Press, 1992). See also the new journal *Ecosystem Health,* which began publication in March 1995.
9. Reed F. Noss and Allen Y. Cooperrider, *Saving Nature's Legacy* (Washington, D.C.: Island Press, 1994).
10. Robert Lawton, personal communication (November, 1996).

Afterword

For many of our students, the day that we spend with Bren Whittaker, whom we introduced at the start of Chapter 9, is the single most important day of our year together, a year in which we visit a tropical rainforest and a temperate old-growth hardwoods forest, in which we see howler monkeys and Plymouth gentians, and in which we meet a great number of talented people who have dedicated their lives to the conservation of biodiversity.

For the last couple of years we have started our day with Bren standing along a dead-end dirt road in northern Vermont. After piling out of our vehicles and pulling on waterproof boots, our students gather to meet Bren. As he has done each year that we have visited him, the first thing he does is ask for each student's name, home, and field of study—and he remembers most of the information he receives. Throughout the day, he will turn to a student and ask how the situation that we are discussing compares with his experience in Alaska or her life in South Carolina. Invariably Bren listens with respect. He does not use the students to make specific points; instead, he hopes to learn as much as he teaches.

After the introduction, Bren speaks briefly and with quiet pride about the conservation importance of the land where we are standing, of how, as Secretary of the Environment for Vermont, he enabled the state to purchase Moose Bog and the surrounding forest. We move off to the bog itself, walking along ground that becomes progressively wetter until we are out of the forest and standing on the floating mat of the bog. We find the footprint of a moose, or perhaps some fresh droppings, giving credence to the site's name. If we were to end our time with Bren at the bog it would be quite special, but after we leave the bog we drive to his home, the 63 acres of land where he and his wife Dorothy have lived since 1959.

The love that Bren holds for his land is deep and complex. This is a man who understands that the land and its biodiversity sustain our species, and that his land nourishes his family in many ways. As a professional forester, pastor in a poor rural community, husband of a farmer in the rocky soil of northern New England, and conservationist, Bren holds many values toward his land and understands the many forms of worth and sustenance that the land offers. Moreover, he does not just view his land in terms of the present. He shows our students the line where the last glacier stopped for a while, some 12,000 years ago, creating an outwash plain like the ones we see on Martha's Vineyard. The edge of the plain is clearly marked where his neighbor's corn field ends, for beyond, the crop would not grow. Later, deep in his forest, we stand around the decayed stump of a great white pine that was cut perhaps a century ago. The tree was so large that a large mound still remains, the remnants of a forest giant that may have begun life before Samuel Champlain visited the New World. We visit the site where a shack once stood, home to a family during the Depression. Although no trace of the building itself remains, the herbs and shrubs at the site indicate the former presence of a house to Bren's practiced eye. Bren shows us another portion of his forest where the trees are all about the same age. The place, he tells us, took the full force of the Great New England Hurricane of 1938, which flattened many forests in New England, including several patches of old growth. The legacy of the storm can be read from the even-aged trees standing at this spot.

Bren's appreciation of his land is not only historical; he also thinks of the future. He also shows us the seven acres of red pine plantation that he planted in 1961, soon after he acquired the property. The pines are reaching marketable size, and he expects that in a few years one of his sons will begin harvesting and selling the trees for use as utility poles. Elsewhere on his land, we find white plastic tubes surrounding slender saplings of white ash and sugar maple. Bren shows us saplings that are not protected, and how they have been repeatedly chewed down by deer. The tubes are one of his attempts to give some saplings a chance to grow tall enough to survive the deer. Bren the forester speaks to us of the market value of these hardwoods; Bren the conservationist has a twinkle in his eye, and we hear in his voice that the real reason he protects these tiny trees is to help re-establish the diverse forest that stood on his land, before the number of deer grew so high.

Until 1995, he always showed our class two neighboring stands: one that he had cleared completely, according to a certain school of forestry practice, and another that he left forested, occasionally cutting specific trees for market or to help other trees grow to a marketable size. Between two of our annual visits, however, the cleared site, which was an impenetrable thicket and produced no potentially marketable trees over the 11 years he had left it, had come under the control of another steward of the land—a family of beavers. The beavers had flooded the thicket, ending Bren's experiment.

Throughout our time together, Bren acknowledges that his land is not merely an historical entity, nor just a storehouse for the future, but that it has great worth in the present. Bren and his wife Dorothy tell us of the small farm that they run, of the tremendous efforts they expend to coax vegetables from rocky soil during a short and unpredictable growing season. They grow much of their own produce here, as well as feeding neighbors and vacationers in the region. The wood that they burn to heat their home during the long winters comes from their land, as does some of the meat and fish for their table.

The most important stop we make on our tour of his land, however, is in a small patch of forest that Bren calls his sanctuary. Here, in the lee of a ridge, stand several tall firs, spruce, and white birch that survived the 1938 hurricane. Here stands a tiny piece of forest that may approach being old growth. This tiny stand is special to Bren, and to each of our students who visit it. Here, the strands of the past, present, and future come together. Here we catch a glimpse of what much of his land might have looked like several hundred years ago. Here we see a legacy that Bren is preserving for his children and grandchildren, and here we draw strength from the mighty trees around us. As a young forester, Bren recognized that this particular spot had much more to offer than mere board feet of lumber, and he left it untouched.

Bren has devoted his life to helping others through his skills as a forester, religious leader, farmer, and conservationist. He brings many perspectives and insights to his own land, and to the landscapes where he, his parishioners, and the people of his state live. He recognizes his responsibilities to use the land and its biodiversity wisely, to provide for those living today—to provide them not only with food and lumber but also with sanctuaries where they can refresh their spirits. But he also recognizes that his stewardship of his land requires a certain sensitivity to the past and regard for the future.

By watching Bren walk his land, and hearing him share the many ways in which he loves and uses its biodiversity, each of us gains new insight into our own values and the worth of biodiversity. We hope that this book, too, will help others gain insight into their values, that they may recognize the many forms of worth to be found in biodiversity, and that they may better protect it.

Bibliography

Alcorn, Janis B. "The CNA Workshop." In *Papua New Guinea Conservation Needs Assessment Volume I*, ed. Janis B. Alcorn. Washington, D. C.: The Biodiversity Support Program, 1993.

Allaby, Michael, ed. *The Concise Oxford Dictionary of Botany*. Oxford: Oxford University Press, 1992.

All Taxa Biological Inventory Protocol, 1993.

Angermeier, Paul L. "Does Biodiversity Include Artificial Diversity?" *Conservation Biology* 8 (1994): 600-02.

Angermeier, Paul L., and James R. Karr. "Biological Integrity Versus Biological Diversity as Policy Directives." *Bioscience* 44 (1994): 690-97.

Ayto, John. *Dictionary of Word Origins*. New York: Arcade Publishing, 1990.

Bailey, Robert G. *Ecosystem Geography*. New York: Springer-Verlag, 1996.

Bazzaz, F. A., and T. W. Sipe. "Physiological Ecology, Disturbance, and Ecosystem Recovery." In *Potentials and Limitation of Ecosystem Analysis*, ed. E. D. Schulze and H. Zwölfer. New York: Springer-Verlag, 1987.

Biodiversity Support Program, Conservation International, The Nature Conservancy, Wildlife Conservation Society, World Resources Institute, and World Wildlife Fund. *A Regional Analysis of Geographic Priorities For Biodiversity Conservation in Latin America and the Caribbean*. Washington, D. C.: Biodiversity Support Program, 1995.

Cody, Martin L. "Diversity, Rarity, and Conservation in Mediterranean-Climate Regions." In *Conservation Biology*, ed. Michael E. Soulé. Sunderland, Massachusetts: Sinauer, 1986.

Compton's Interactive Encyclopedia. Copyright © 1994, 1995 Compton's NewMedia, Inc.

Conant, Roger, and Joseph T. Collins. *A Field Guide to Reptiles and Amphibians*. 3rd edition. Boston: Houghton Mifflin, 1991.

Convention on Biological Diversity. 1992.

Convention on International Trade in Endangered Species of Wild Fauna and Flora.

Corn, M. Lynne. *Ecosystems, Biomes, and Watersheds: Definitions and Use*. Congressional Research Service, 93-655 ENR, 1993.

Costanza, Robert, Bryan G. Norton, and Benjamin D. Haskell. *Ecosystem Health*. Washington, D. C.: Island Press, 1992.

Cox, C. Barry, and Peter D. Moore. *Biogeography*. 5th edition. Oxford: Blackwell Scientific, 1993.

Daily, G. C., and P. R. Ehrlich. Nocturnality and Species Survival. *Proceedings of the National Academy of Sciences*, 93 (1996): 11709–11712.

Daugherty, C. H., A. Cree, J. M. Hay, and M. B. Thompson. "Neglected Taxonomy and Continuing Extinctions of Tuatara (*Sphenodon*)" *Nature* 247 (1990): 177-79.

Davis, M. B. "Quaternary History and the Stability of Forest Communities." In *Forest Succession: Concepts and Application*, ed. D. C. West, H. H. Shugart, and D. B. Botkin. New York: Springer-Verlag, 1981: 132–153.

Davis, M. B. "Climatic Instability, Time Lags, and Community Disequilibrium." In *Community Ecology*, ed. J. Diamond and T. J. Case. New York: Harper & Row, 1986.

DeSylva, B. G. (words) and Louis Silvers (music). "April Showers." Miami, Florida: Warner Bros., Inc., 1921.

Dworkin, R. M. *Taking Rights Seriously*. Cambridge, Massachusetts: Harvard University Press, 1977.

Ehrenfeld, David. *The Arrogance of Humanism*. Oxford: Oxford University Press, 1978.

Ehrenfeld, David. "The Management of Diversity: A Conservation Paradox." In *Ecology, Economics, Ethics: The Broken Circle*, ed. F. Herbert Bormann and Stephen R. Kellert. New Haven, Connecticut: Yale University Press, 1991.

Ehrlich, Paul R., David S. Dobkin, and Darryl Wheye. *The Birder's Handbook*. New York: Simon & Schuster, 1988.

Ereshefsky, Marc, ed. *The Units of Evolution: Essays on the Nature of Species*. Cambridge, Massachusetts: MIT Press, 1992.

Erwin, Terry L. "Tropical Forests: Their Richness in Coleoptera and Other Arthropod Species." *The Coleopterists Bulletin* 36 (1982): 74-75.

Erwin, Terry L. "An Evolutionary Basis for Conservation Strategies." *Science* 253 (1991): 750-52.

Erwin, Terry L. "How Many Species Are There?: Revisited." *Conservation Biology* 5 (1991): 330-333.

Faith, Daniel P. "Phylogenetic Pattern and the Quantification of Organismal Biodiversity." *Philosophical Transactions of the Royal Society London* B 345 (1994).

Fiedler, Peggy L., and Subodh K. Jain, eds. *Conservation Biology: The Theory and Practice of Nature Conservation Preservation and Management*. New York: Chapman and Hall, 1992.

Franklin, J. F., and R. T. T. Forman. "Creating Landscape Patterns by Forest Cutting: Ecological Consequences and Principles." *Landscape Ecology* 1 (1987): 5-18.

Frosst, L. C. "The Study of *Ranunculus ophioglossoides* and Its Successful Conservation at the Badgeworth Nature Reserve, Gloustershire." In *Rare Plant Conservation*, ed. Hugh Synge. New York: Wiley, 1991.

Gaston, K. J. *Biodiversity: A Biology of Numbers and Difference*. Oxford: Blackwell Science, 1996.

Harper, J. L., and D. L. Hawksworth. Preface to "Biodiversity: Measurement and Estimation." *Philosophical Transactions of the Royal Society London B* 345 (1994): 5-12.

Hartshorn, G.S. "Plants." In *Costa Rican Natural History*, ed. Daniel H. Janzen. Chicago: University of Chicago Press, 1983.

Hesiod, *Works and Days*. In *Hesiod and Theognis*, trans. Dorothea Wender. London: Penguin Books, 1973.

Hölldobler, Bert, and Edward O. Wilson. *The Ants*. Cambridge, Massachusetts: Harvard University Press, 1990.

Hunter, Malcolm L., Jr. *Fundamentals of Conservation Biology*. Cambridge, Massachusetts: Blackwell Science, 1996.

INBio. Costa Rican National Institute of Biodiversity Web site. http://www.inbio.ac.cr/ATBI/ATBI.html on July 29, 1996.

International Council for Bird Preservation. *Putting Biodiversity on the Map: Priority Areas for Global Conservation*. Cambridge: International Council for Bird Preservation, 1992.

International Institute for Sustainable Development. Linkages Web site. http://www.iisd.ca/linkages/vol09/0918001e.html on December 5, 1996. http://www.inbio.ac.cr/ATBI/ATBILetter.html on June 13 1997; http://www.inbio.accr/ATBI/ATBINews2.html on June 13 1997.

IUCN. *World Conservation Strategy*. Gland, Switzerland: IUCN, 1980.

IUCN, UNEP, WWF, and World Commission on Environment and Development. *Our Common Future*. Oxford: Oxford University Press, 1987.

Janzen, D. J. "A South-North Perspective on Science in the Management, Use, and Economic Development of Biodiversity." In *Conservation of Biodiversity for Sustainable Development*, ed. O. T. Sandlund et al. Oslo: Scandinavian University Press, 1992.

Johnsgard, Paul. *Cranes of the World*. Bloomington: Indiana University Press, 1983.

Johnson, Nels C. *Biodiversity in the Balance: Approaches to Setting Geographic Conservation Priorities*. Washington, D. C.: Biodiversity Support Program, 1995.

Joyce, Christopher. "Taxol: Search for a Cancer Drug." *BioScience* 43 (1993): 133.

Karr, J. R., and D. R. Dudley. "Ecological Perspective on Water Quality Goals." *Environmental Management* 5 (1981): 55-68.

Keeton, William T., and James L. Gould. *Biological Science*. 5th edition. New York: W. W. Norton, 1993.

Keller, E. F., and E. A. Lloyd. *Keywords in Evolutionary Biology*. Cambridge, Massachusetts: Harvard University Press, 1992.

Kellert, Stephen R. "The Biological Basis for Human Values of Nature." In *The Biophilia Hypothesis*, ed. Stephen R. Kellert and Edward O. Wilson. Washington, D.C.: Island Press, 1992.

Kellert, Stephen R., and Edward O. Wilson, eds. *The Biophilia Hypothesis*. Washington, D. C.: Island Press, 1992.

Kellert, Stephen R. *The Value of Life*. Washington, D. C.: Island Press, 1995.

Latham, Roger Earl, and Robert E. Ricklefs. "Continental Comparisons of Temperate-Zone Tree Species Diversity." In *Species Diversity in Ecological Communities*, ed. Robert E. Ricklefs and Dolph Schluter. Chicago: Chicago University Press, 1993.

Leopold, Aldo. *Round River*. New York: Oxford University Press, 1993.

Lewin, Roger. *Human Evolution: An Illustrated Introduction*. Cambridge, Massachusetts: Blackwell Science, 1993.

Lincoln, Roger J., and G. A. Boxshall. *The Cambridge Illustrated Dictionary of Natural History*. Cambridge, United Kingdom: Cambridge University Press, 1987.

Maddison, W.P. and D.R. Maddison. *Manual to MacClade*, version 3. Sunderland, Massachusetts: Sinauer Associates, 1992.

Magurran, Anne. *Ecological Diversity and Its Measurement*. Princeton, New Jersey: Princeton University Press, 1988.

Mangelsdorf, Paul C. *Corn: Its Origin Evolution and Improvement*. Cambridge, Massachusetts: Harvard University Press, 1974.

Margulis, Lynn, and Karlene V. Schwartz. *Five Kingdoms*. 2nd edition. New York: W.H. Freeman, 1988.

Martin, Paul S., and Richard G. Klein, eds. *Quaternary Extinctions*. Tucson: University of Arizona Press, 1984.

May, Robert M. "Levels of Organization in Ecology." In *Ecological Concepts*, ed. J. M. Cherrett. Oxford: Blackwell Scientific Publications, 1989.

May, R. M. "Taxonomy as Destiny." *Nature* 347, no. 6289 (1990): 129-30.

May, Robert M. "Conceptual Aspects of the Quantification of the Extent of Biological Diversity." In *Biodiversity: Measurement and Estimation*, ed. D. L. Hawksworth. London: Chapman & Hall, 1994.

Mayr, Ernst. "Speciation Phenomena in Birds." *American Naturalist* 74 (1940): 249-278.

Mayr, Ernst. *Animal Species and Evolution*. Cambridge, Massachusetts: Harvard University Press, 1963.

McDade, Lucinda A. "Hybridization and Phylogenetics." In *Experimental and Molecular Approaches to Plant Biosystematics*, ed. P. Hoch and A. G. Stephenson. St. Louis, Missouri: Missouri Botanical Garden, 1995.

McNeely, Jeffrey A., and Kenton R. Miller, Walter V. Reid, Russell A. Mittermeier, and Timothy B. Werner. *Conserving the World's Biological Diversity*. Gland, Switzerland: IUCN, 1990.

Meffe, Gary K., and Ronald C. Carroll. *Principles of Conservation Biology*. Sunderland, Massachusetts: Sinauer, 1994.

Menand, Louis. "Diversity." In *Critical Terms for Literary Study*, ed. F. Lentricchia and T. McLaughlin. Chicago: University of Chicago Press, 1995.

Meyer, Axel, Thomas D. Kocher, Pereti Basasibwaki, and Allan C. Wilson. "Monophyletic Origin of Lake Victoria Cichlid Fishes Suggested by Mitochondrial DNA Sequences." *Nature* 347 (1990): 550-553.

Mittermeier, Russell A. "Primate Diversity and the Tropical Forest." In *Biodiversity*, ed. Edward O. Wilson. Washington, D. C.: National Academy Press, 1988.

Monmonier, Mark. *How to Lie with Maps*. Chicago: Chicago University Press, 1991.

Moyle, Peter B., and Joseph J. Cech, Jr. *Fishes: An Introduction to Ichthyology*. 3rd edition. New Jersey: Prentice Hall, 1996.

Myers, Norman. *The Primary Source*. New York: W. W. Norton, 1984.

Nagy, Gregory. "Theognis of Megara: A Poet's Vision of his City." In *Theognis of Megara: Poetry and the Polis*, ed. T. J. Figueira and Gregory Nagy. Baltimore, Maryland: John Hopkins University Press, 1985.

Nagy, Gregory. *Pindar's Homer*. Revised Paperback Edition. Baltimore, Maryland: Johns Hopkins Press, 1994.

Needham, James G. and Minter J. Westfall, Jr. *A Manual of the Dragonflies of North America*. Berkley: University of California Press, 1954.

Norton, Bryan G. *Why Preserve Natural Variety?* Princeton, New Jersey: Princeton University Press, 1987.

Norton, Bryan G., and Robert Ulanowicz. "Scale and Biodiversity Policy: A Hierarchical Approach." *Ambio* 21 (1992): 244-249.

Norton, Bryan G. "On What We Should Save: The Role of Culture in Determining Conservation Targets." In *Systematics and Conservation Evaluation*, ed. P. L. Forey, C. J. Humphries, and R. I. Vane-Wright. Oxford: Oxford University Press, 1994.

Noss, R. F. "Indicators for Monitoring Biodiversity: A Hierarchical Approach." *Conservation Biology* 4, no. 4 (1990): 355-364.

Noss, Reed F., and Allen Y. Cooperrider. *Saving Nature's Legacy*. Washington, D. C.: Island Press, 1994.

Orians, Gordon. "Endangered at What Level?" *Ecological Applications* 3, no. 2 (1993): 206-208.

Palmer, Thomas. "The Case For Human Beings." *The Atlantic Monthly* 269 (1992): 83–88.

Pearce, David, and Dominic Moran. *The Economic Value of Biodiversity*. London: Earthscan Publication Limited, 1994.

Pielou, E. C. *After the Ice Age*. Chicago: University of Chicago Press, 1991.

Pollard, John. *Bird in Greek Life and Myth*. Plymouth, England: Thames and Hudson, 1977.

Prance, G. T. "Discussion." In *Vicariance Biogeography: A Critique*, ed. D. Nelson and D. E. Rosen. New York: Columbia University Press, 1981.

Prance, G.T. "Forest Refuges: Evidence from Woody Angiosperms." In *Biological Diversification in the Tropics*, ed. G. T. Prance. New York: Columbia University Press, 1982.

Prange, Hartwig. "Crane." In *Birds in Europe: Their Conservation Status*, ed. G. M. Tucker and M. F. Heath. Cambridge, United Kingdom: BirdLife International, 1994.

Prendergast, J. R., R. M. Quinn, J. H. Lawton, B. C. Evershame, and D. W. Gibbons. "Rare Species, the Coincidence of Diversity Hotspots and Conservation Strategies." *Nature* 365 (1993): 335-337.

Pressey, R. L., C. J. Humphries, C. R. Margules, R. I. Vane-Wright, and P. H. Williams. "Beyond Opportunisms: Key Principles for Systematic Reserve Selection." *Trends in Ecology and Evolution* 8 (1993): 124-128.

Preston, F. W. "Diversity and Stability in the Biological World." In *Diversity and Stability in Ecological Systems*. Brookhaven Symposia in Biology, No. 22., 1969.

Primack, Richard B. *Essentials of Conservation Biology*. Sunderland, Massachusetts: Sinauer, 1993.

Raeburn, Paul. *The Last Harvest*. New York: Simon and Schuster, 1995.

Reid, Walter V., and Kenton R. Miller. *Keeping Options Alive*. Washington, D. C.: World Resources Institute, 1989.

Richards, A. J. *Plant Breeding Systems*. London, Allen and Unwin, 1986.

Ridley, Mark. *Evolution*. Boston, Massachusetts: Blackwell Scientific Publications, 1993.

Rojas, Martha. "The Species Problem and Conservation: What Are We Trying to Protect?" *Conservation Biology* 6 (1992): 170-178.

Rolston, Holmes. *Conserving Natural Value*, New York: Columbia University Press, 1994.

Rzach, Aloisius, *Hesiodi Carmina*. Lipsiae in Aedibus B.G. Tebneri, 1902.

Schluter, Dolph, and Robert E. Ricklefs. "Species Diversity: An Introduction to the Problem." In *Species Diversity in Ecological Communities*, ed. Robert E. Ricklefs, and Dolph Schluter. Chicago: University of Chicago Press, 1993.

Scott J. Michael, Blair Csuti, James D. Jacobi, and John E. Estes. "Species Richness." *BioScience* 37 (1987): 782–788.

Sen, A. K. *Inequality Re-examined*. Cambridge, Massachusetts: Harvard University Press, 1992.

Shaffer, Mark L. "Minimum Population Sizes for Species Conservation." *Bioscience* 31 (1981): 131-134.

Shrader-Frechette, K. S., and McCoy, eds. *Method in Ecology*, Cambridge: Cambridge University Press, 1993.

Speirs, J. M. *Birds of Ontario*. Toronto: Natural Heritage/Natural History, 1985.

Spilliaert, R., G. Vikingsson, H. Arnason, A. Palsdottir, J. Sigurjonsson, and A. Arnason. "Species Hybridization Between a Female Blue Whale (*Balaenoptera musculus*) and a Male Fin Whale (*B. physalus*): Molecular and Morphological Documentation." *The Journal of Heredity* 82 (1991): 269-275.

Steadman, David. "Human-Caused Extinction of Birds." In *Biodiversity II*, ed. M. L. Reka-Kudla, D. E. Wilson, and E. O. Wilson. Washington, D. C.:, Joseph Henry Press, 1997.

Stiles, F. Gary, and Alexander F. Skutch. *A Guide to the Birds of Costa Rica*. Ithaca, New York: Cornell University Press, 1989.

Stork, Nigel E. "Insect Diversity: Facts, Fiction and Speculation." *Biological Journal of the Linnean Society*, 35 (1988): 321-337.

Swartzendruber, J. Fred. "Executive Summary." In *Papua New Guinea Conservation Needs Assessment Volume I*, ed. Janis B. Alcorn. Washington, D. C.: The Biodiversity Support Program, 1993.

Tennessee Valley Authority *vs.* Hiram Hill, *et al.* United States Supreme Court 437 U.S. 153, 1978.

Uhl, N. W., and J. Dransfield. *Genera Palmarum*. Lawrence, Kansas: Allen Press, 1987.

United Nations Environment Programme. *Global Biodiversity Assessment*. Cambridge: Cambridge University Press, 1995.

United Nations Environment Programme. Web site: http://www.unep.ch/bio/ratif.html on April 5, 1995.

United States Department of Agriculture. *Apples: Production Estimates and Important Commercial Districts and Varieties*. USDA Bulletin No. 485, 1917.

United States Department of Agriculture. *Sylvics of North America*. Vol. 1. Washington, D. C.: United States Government Printing Office, 1990.

United States Government Accounting Office *Endangered Species Act: Types and Number of Implementing Action.* GAO/RCED-92-131BR, 1992.

United States Government Accounting Office *Ecosystem Management.* GAO/RCED-94-111, 1994.

U.S. Office of Technology Assessment. *Federal Technology Transfer and the Human Genome Project.* Washington, D.C.: United States Government Printing Office, 1995.

U.S. Office of Technological Assessment. *Technologies to Maintain Biological Diversity.* Washington, D.C.: United States Government Printing Office, 1987.

Usher, Michael B. "Wildlife Conservation Evaluation: Attributes, Criteria and Values." In *Wildlife Conservation Evaluation,* ed. Michael B. Usher. London: Chapman and Hall, 1986.

Walker, D. "Diversity and Stability." In *Ecological Concepts,* ed. J. M. Cherrett. Oxford: Blackwell Scientific Publications, 1989.

Wayne, R. K., and J. L. Gittleman "The Problematic Red Wolf." *Scientific American* (July 1995): 36–39.

Weitzman, Martin. "What to Preserve: An Application of Diversity Theory to Crane Conservation." *Quarterly Journal of Economics* 108, no. 1 (1993): 157-84.

White, Gilbert. *The Natural History and Antiquities of Selbourne.* Harmondsworth, England: Penguin Books, 1977.

White, Michael J. D. *Modes of Speciation.* San Francisco: W. H. Freeman and Company, 1978.

Whitmore, T. C. *An Introduction to Tropical Rain Forests.* Oxford: Oxford University Press, 1990.

Whittaker, R. H. "Vegetation of the Great Smokey Mountains." *Ecological Monographs* 23 (1956): 41–78.

Williams, P.H., R. I. Vane-Wright, and C. J. Humphries. "Measuring Biodiversity for Choosing Conservation Areas." In *Hymenoptera and Biodiversity,* ed. John LaSalle and Ian D. Gauld. Wallingford, United Kingdom: CAB International, 1993.

Williams, R. *Keywords: A Vocabulary of Culture and Society.* New York: Oxford University Press, 1976.

Wilson, Edward O. *Biophilia.* Cambridge, Massachusetts: Harvard University Press, 1984.

Wilson, Edward O., ed. *BioDiversity.* Washington, D.C.: National Academy Press, 1988.

Wilson, Edward O. *The Diversity of Life.* Cambridge, Massachusetts: Harvard University Press, 1992.

Wirth, Timothy E. "The Road From Rio — Defining a New World Order." *Colorado Journal of International Environmental Law and Policy* 4 (1993): 37-44.

World Conservation Monitoring Centre. *Global Biodiversity: Status of the Earth's Living Resources.* London: Chapman & Hall, 1992.

World Resources Institute. *World Resources 1994-95.* Oxford: Oxford University Press, 1994.

World Resources Institute. *World Resources 1996-97.* New York: Oxford University Press, 1996.

Wynne, Peter. *Apples.* New York: Hawthorn Books, 1975.

Yoon, Carol K. "Counting Creatures Great and Small." *Science* 260 (1993): 620-22.

Zeleny, L. *The Blue Bird—How You Can Help Its Fight for Survival.* Bloomington: Indiana University Press, 1976.

Zimmerman, Elwood C. *Insects of Hawaii.* Honolulu: University of Hawaii Press, 1948.

Zonnefeld, I. S., and R. T. T. Forman. *Changing Landscapes: An Ecological Perspective.* New York: Springer-Verlag, 1990

Zug, George R. *Herpetology.* San Diego, California: Academic Press, 1993.

Index

Note: Page numbers followed by an *f, t, b,* or *n* indicate figures, tables, boxes, or footnotes, respectively; *pl.* indicates color plate number.